Chemical Analysis of Contaminated Land

Analytical Chemistry Series

Series Editors: John M. Chalmers and Alan J. Handley

A series which presents the current state of the art in chosen sectors of analytical chemistry. Written at professional and reference level, it is directed at analytical chemists, environmental scientists, food scientists, pharmaceutical scientists, earth scientists, petrochemists and polymer chemists. Each volume in the series provides an accessible source of information on the essential principles, instrumentation, methodology and applications of a particular analytical technique.

Titles in the series:

Inductively Coupled Plasma Spectrometry and its Applications
Edited by S.J. Hill

Extraction Methods in Organic Analysis
Edited by A.J. Handley

Design and Analysis in Chemical Research
Edited by R.L. Tranter

Spectroscopy in Process Analysis
Edited by J.M. Chalmers

Gas Chromatographic Techniques and Applications
Edited by A.J. Handley and E.R Adlard

Atomic Spectroscopy in Elemental Analysis
Edited by M. Cullen

Pharmaceutical Analysis
Edited by D.C. Lee and M. Webb

Chemical Analysis of Contaminated Land
Edited by K.C. Thompson and C.P. Nathanail

Chemical Analysis of Contaminated Land

Edited by

K. CLIVE THOMPSON
ALcontrol Laboratories Rotherham, UK

and

C. PAUL NATHANAIL
Land Quality Management
University of Nottingham, UK

Blackwell
Publishing

CRC Press

Editorial Offices:
9600 Garsington Road, Oxford OX4 2DQ, UK
  Tel: +44 (0) 1865 776868
108 Cowley Road, Oxford OX4 1JF, UK
  Tel: +44 (0)1865 791100
Blackwell Munksgaard, 1 Rosenørns Allé,
P.O. Box 227, DK-1502 Copenhagen V,
Denmark
  Tel: +45 77 33 33 33
Blackwell Publishing Asia Pty Ltd,
550 Swanston Street, Carlton South, Victoria
3053, Australia
  Tel: +61 (0)3 9347 0300
Blackwell Verlag, Kurfürstendamm 57,
10707 Berlin, Germany
  Tel: +49 (0)30 32 79 060
Blackwell Publishing, 10 rue Casimir
Delavigne, 75006 Paris, France
  Tel: + 33 1 53 10 33 10

Published in the USA and Canada (only) by
CRC Press LLC
2000 Corporate Blvd., N.W.
Boca Raton, FL 33431, USA
Orders from the USA and Canada (only) to
CRC Press LLC

USA and Canada only:
ISBN 0–8493–2810–1

This book contains information obtained from
authentic and highly regarded sources. Reprinted
material is quoted with permission, and sources
are indicated. Reasonable efforts have been
made to publish reliable data and information,
but the author and the publisher cannot assume
responsibility for the validity of all materials
or for the consequences of their use.

Trademark notice: Product or corporate names
may be trademarks or registered trademarks, and
are used only for identification and explanation,
without intent to infringe.

First published 2003

A catalogue record for this title is available
from the British Library

ISBN 1–84127–334–1
Originated as Sheffield Academic Press

Library of Congress
Cataloging-in-Publication Data
is available

Set in 10.5/12 pt Times
by Integra Software Services Pvt Ltd,
Pondicherry, India
Printed and bound in Great Britain by
MPG Books Ltd, Bodmin, Cornwall

For further information on
Blackwell Publishing, visit our website:
www.blackwellpublishing.com

Contents

4 Metal analysis 64
PATRICK THOMAS

5 Analysis of inorganic parameters 99
GEORGE E. RAYMENT, ROSS SADLER, ANDREW CRAIG,
BARRY NOLLER and BARRY CHISWELL

6   Petroleum hydrocarbons and polyaromatic hydrocarbons   132
JIM FARRELL-JONES

7   Volatile organic compounds   177
SUE OWEN and PETER WHITTLE

Contributors

Dr Mark Allen   British Geological Survey, Nicker Hill, Keyworth, Nottingham NG12 5GG, UK

Dr Rudolf Braun   Contaminated Land Management, IFA-Tulln, Environmental Biotechnology, Konrad Lorenz Strasse 20, A-3430 Tulln, Austria

Professor Barry Chiswell   National Research Centre for Environmental Toxicology, The University of Queensland, 39 Kessels Road, Coopers Plains, Queensland, 4108, Australia

Mr Andrew Craig   ALcontrol Laboratories, Uxbridge Laboratory, Brunel Science Park, Kingston Lane, Uxbridge, Middlesex UB8 3PQ, UK

Mr Paul Frintrop   Smedinghuis, Zuiderwagenplein 2, 8224 AD Lelystad, PO Box 17, 8200 AA Lelystad, The Netherlands

Mr Joop Harmsen   Alterra Institute, PO Box 47, 6700 AA Wageningen, The Netherlands

Ms Leslie Heasman   MJ Carter Associates, Station House, Long Street, Atherstone, Warwickshire CV9 1BH, UK

Dr Doris Hirmann   Contaminated Land Management, IFA-Tulln, Environmental Biotechnology, Konrad Lorenz Strasse 20, A-3430 Tulln, Austria

Mr Jim Farrell-Jones   ALcontrol, Chester Laboratory, Chester Street, Saltney, Chester CH4 8RD, UK

Dr Andreas P. Loibner   Contaminated Land Management, IFA-Tulln, Environmental Biotechnology, Konrad Lorenz Strasse 20, A-3430 Tulln, Austria

Dr C. Paul Nathanail   Land Quality Management, School of Chemical, Environmental and Mining Engineering, The University of Nottingham, University Park, Nottingham NG7 2RD, UK

Professor Barry Noller   National Research Centre for Environmental Toxicology, The University of Queensland, 39 Kessels Road, Coopers Plains, Queensland 4108, Australia

Ms Sue Owen   AES, Northumberland Dock Road, Wallsend, Tyne and Wear, NE28 0QD, UK

Mr George E. Rayment   Queensland Department of Natural Resources and Mines, Block B, 80 Meiers Road, Indooroopilly, QLD 4068, Australia

Dr Ross Sadler   Queensland Health Scientific Services, PO Box 594, Archerfield, Queensland, Australia

Dr Oliver H.J. Szolar   Contaminated Land Management, IFA-Tulln, Environmental Biotechnology, Konrad Lorenz Strasse 22, A-3430 Tulln, Austria

Dr Patrick Thomas   Department of Water and Environment, Institut Pasteur de Lille, 1 rue de Professeur Calmette, BP 245, 59019 Lille cedex, France

Professor K. Clive Thompson   ALcontrol Laboratories, Templeborough House, Mill Close, Rotherham S60 1BZ, UK

Dr David Westwood   Environment Agency, Wheatcroft Office Park, Landmere Lane, Edwalton, Nottingham NG12 4DG, UK

Dr Peter Whittle   6 Marian Drive, Rainhill, Prescot, Merseyside L35 0NB, UK

Preface

This book sets out to provide a description of the chemical analysis of potentially contaminated land for all those involved in risk assessment. It is not intended as a recipe book of analytical methods. Indeed, with the wide range of techniques and options available, it would be impracticable and unnecessary to document all potential methods, and methods can rapidly become superseded. The book aims to assist in specifying appropriate analyses, relevant strategies for carrying out analyses and methods of interpreting results within the new risk-based legislative framework for contaminated land. Risk assessors with no formal training in, or knowledge of, chemical analysis require an appreciation of the strengths and limitations of such analyses if they are to use analytical results in an intelligent and informed way.

The book is directed equally at the analytical chemist and the risk assessor (environmental scientist or engineer) responsible for commissioning analyses of potentially contaminated soil or water samples. It is written in a way that should prove helpful to both new and experienced practitioners. Other stakeholders in contaminated land management, such as environmental consultants, developers, land owners, legal advisers, insurers, local government and environmental protection agencies, will also find the volume of assistance. The book will provide analytical laboratories with an insight into the concerns and requirements of risk assessors.

The book comprises ten chapters, distilling the expertise and experience of the authors in the analysis of contaminated land. It commences with an introductory chapter setting the context within which risk assessors commission analyses of potentially contaminated soil and water samples. Results should be fit for purpose. Analysis is simply a means of reducing analytical uncertainty to an acceptable level. Analytical error is minor when compared to natural variability and sampling error. Chapter 2 highlights the difficulty in developing robust methods of analysis that cover all the soil matrices likely to be encountered. It stresses the need for comprehensive method validation, which is necessary to ensure that fit-for-purpose results are obtained. Chapter 3 tackles the problem of preparing the submitted samples for analysis. If inappropriate sample homogenisation and initial preparation protocols are employed, then, no matter how accurate the subsequent analysis of the incorrectly prepared sub-samples, the results will not reflect conditions in the field and will therefore be of little use to the risk assessor. The following five chapters cover the analysis of metals, including a brief discussion on speciation analysis (Chapter 4), the analysis of inorganic parameters such as cyanide, pyrites, total sulfur and asbestos (the latter being a commonly overlooked area) (Chapter 5), petroleum hydrocarbons, including polyaromatic hydrocarbons (Chapter 6), volatile organic compounds (Chapter 7) and non-halogenated organic compounds, including semi-volatile organic compounds (SVOCs) (Chapter 8). Analysis of organic contaminants is a particularly complex area. It is difficult to obtain accurate data for many organic parameters from many soil matrices, and these last three chapters covering organic analysis issues, attempt to outline strategies for obtaining results which are fit for purpose.

Chapter 9 deals with leaching tests. These tests are used to assess the mobility of contaminants and their rate of release from a solid form or soil matrix into water. Leaching tests are a common component of groundwater risk assessment, and so it is important to appreciate that all leaching methods are empirical and that the results are defined by the leaching method. Small changes (e.g. in the temperature or shaking regime) can have a very significant effect upon the results.

Chapter 10 describes the use of toxicity tests in ecological assessment and toxicity screening of potentially contaminated land. Ecotoxicity can be assessed only by the application of biological methods, whereas chemical analysis determines concentrations of defined chemicals from which toxic effects may be deduced or predicted. Assessment criteria for contaminated land, sediments, leachates or groundwater should ideally be derived from observation of biological effects. Ecotoxicity results give a holistic answer relating to the whole sample. This is a relatively new area in contaminated land analysis and one which is confidently predicted to grow significantly over the coming years.

We hope that we have produced a book that will prove useful to both laboratories working on the chemical analysis of potentially contaminated land and their direct and indirect customers.

K. Clive Thompson
C. Paul Nathanail

1 The risk assessor as the customer

K. Clive Thompson and C. Paul Nathanail

1.1 Analysis issues

Many countries have adopted a risk-based approach to assessing and tracking the historic contamination that is a legacy of past industrial practice. The approach involves identifying, and then linkages, unacceptable contaminant source, pathway and receptor *pollutant increases*. Sampling and analysis of potentially contaminated sites provide essential information to risk assessors. Sampling aspects are not covered in this book, as there are relevant comprehensive texts covering this key area (DoE 1994; Ferguson *et al.* 1998; Ramsey 1998; BS 5930:1999; Environment Agency 2000a,b; BS 10175:2001; ISODIS 10381:2002). However, there are few texts covering the analytical aspects of site investigations.

It is essential to appreciate that the reliability of any analysis can only relate to the sample submitted to the laboratory. If unfit-for-purpose samples (e.g. inappropriately sampled and/or incorrectly preserved) are submitted to a laboratory, the results produced by the laboratory (no matter how accurate) will also be unfit for the intended purpose of assessing the risks associated with the site. Thus, it is very important that the laboratory is supplied with correctly taken and suitably preserved samples and that these samples are representative of the location and depth from where they were taken. In addition, the samples should reflect the portion of the soil that is involved in a given pathway.

There are many objectives for a site investigation (Barr *et al.* 2002). These may include the following:

1. To refine and clarify a conceptual model.
2. To inform risk estimation.
3. To provide relevant data for the selection or design of remedial works.
4. To benchmark the contamination status.
5. To provide information for the assessment of potential future liabilities of a site.
6. To verify remediation has achieved its clean-up targets.

These objectives are supported by the selection of appropriate relevant analyses. Effective contaminated land management requires an integrated approach from a multidisciplinary range of scientists and other professionals. There is always a finite budget and limited time, but there is a wide range of analysis strategies. It is always advisable to involve the analytical laboratory at an early stage in this process to discuss the various analysis strategies and their associated performance and costs. The major aim is to determine the most cost-effective analysis strategy to achieve the objectives. Part of the strategy will focus on how the chemical composition of potentially contaminated media, and the fate and transport properties of contaminants, will be determined. While modelling or use of literature values has a role to play, the late Colin Ferguson's adage 'measure if you can, model if you must' remains true.

There are three main types of analysis:

1. Targeted analysis.
2. Screening analysis.
3. Ecotoxicological analysis.

Targeted analysis (e.g. the detection and determination of concentrations of specific toxic metals, total cyanide, PAHs etc.) investigates the presence of specific substances, and generally the significance of the results is relatively straightforward to interpret. Screening analysis, usually significantly less costly than targeted analysis, investigates groups of compounds (e.g. toluene extractable matter; mineral oil content (by infrared spectrometry); sum of diesel range organics; sum of polyaromatic hydrocarbons; sum of volatile organic compounds (total VOCs etc.)). Screening analysis is used to assess whether, and to what extent, further laboratory analysis is required. These tests are used to indicate hotspots of pollutants, but the significance of the results is more difficult to interpret than that of targeted analysis (e.g. mineral oils by infrared give very little useful information on the nature of hydrocarbons present (Heath *et al.* 1993)). An ecotoxicological approach (see Chapter 10) gives a holistic indication of all the toxic contaminants present in the sample expressed as an adverse effect on a specified organism or battery of organisms. Again, it does not identify or quantify toxicants present in the sample. This form of analysis can be considered to be a more generalised form of screening analysis where the analysis parameter is usually acute toxicity (although there are some chronic ecotoxicity tests) (see Table 1.1).

Targeted analysis is normally used to determine substances which are present in unacceptable concentrations. Screening analysis is normally used to minimise the amount of targeted analysis by highlighting *hotspots* and eliminating those parts of the site where screening analysis gave negative (*nothing present*) results. Many chemical screening tests no matter how accurate, fortunately tend to have a positive bias leading to false positive results (e.g. toluene extractable matter in peaty soils overestimates the presence of organic contaminants). This can result in further, unnecessary, analysis but ensures a fail-safe analytical strategy. There is a general move away from screening tests such as TPH and total PAH that are insufficiently specific.

Table 1.1 The three main types of contaminated land analysis

Targeted analysis	Screening analysis	Ecotox analysis
Specified toxic metals	Aliphatics; aromatics and non-specific organics by Iatroscan™	*Daphnia magna* inhibition on a leachate
Easily liberated cyanide	Toluene extractable matter	Algal inhibition on a leachate
Total cyanide	Gross alpha and beta radioactivity	Root elongation test
Individual phenols by HPLC	Total monohydric phenols by distillation with colorimetric end point	Ostracod chronic ecotoxicity test on a soil sample
Diesel range organics by GC	Total hydrocarbons by infrared spectrometry	Bait lamina tests
EPA 16 PAHs by GC–MS	Total PAH using GC–FID	Litterbug test
Organochlorine pesticides	AOX or EOX analysis for total organic chlorine	Biosensors
Elemental sulfur; water soluble sulfate and sulfide	Total sulfur	
Chromium VI Physiologically based extraction test		

Ecotoxicological tests (see Chapter 10), normally carried out after leaching the soil with an appropriate leaching reagent, give an indication of the overall toxicity of the sample towards the given test organism. They are considered relevant in detecting toxic hotspots and confirming whether bioremediation has significantly reduced the overall toxicity in the soil. They can highlight problems from unsuspected toxic contaminants that have been missed by targeted or screening analysis. It is only recently that the cost of some relevant ecotoxicological tests has significantly fallen to levels that start to make this approach commercially viable (Persoone *et al.* 2000; Finnamore *et al.* 2002). However, the authors firmly believe that ecotoxicity tests are unlikely to replace targeted or screening chemical analysis (e.g. most acute ecotoxicity leachate tests do not respond to high levels of carcinogenic PAHs as these substances are virtually insoluble in water and do not express high acute toxicity). Ecotox tests are considered complementary to chemical analysis. They can help to improve the reliability of the overall risk assessment.

This book attempts to cover chemical and ecotoxicological analysis related to routine contaminated land investigations. It does not cover analysis related to research or specialist one-off project type investigations. The following chapter deals with soil analysis method requirements, how methods should be validated and the need for all methods to meet clearly defined performance requirements. It also covers quality assurance/quality control aspects. Chapter 3 covers the key, and problematic area of sample homogenisation and the initial sample preparation. Chapter 4 covers the analysis of metals and elemental analysis. The vast majority of elemental analysis is related to *total* analysis, in which all forms of the substances of interest (e.g. tin and mercury) are expected and detected by the analysis and a single total concentration reported for each element. However, the toxicity of the various species of a given element (e.g. inorganic and tri-butyl tin species; inorganic and methyl mercury species, Cr^{III} and Cr^{VI}) can vary by orders of magnitude. Consequently speciation analysis, where individual forms of the substance of interest are analysed, is gaining importance. Chapter 5 deals with the analysis of inorganic parameters (e.g. electrical conductivity, pH, chloride, redox, cyanide, cyanate, thiocyanate, water soluble boron, various sulfur species including sulfide and pyrites). This chapter also covers asbestos. There are increasing concerns over the long-term effects of asbestos and the assessment of potential asbestos contamination in soil samples is of considerable importance to owners of many potentially contaminated land sites.

Chapter 6 covers the analysis of petroleum hydrocarbons including polyaromatic hydrocarbons (PAHs). There is some disagreement on how this analysis should be carried out and results interpreted, and this chapter attempts to deal with these complex issues. The advantages and disadvantages of targeted and screening analyses for these parameters are also covered. Chapter 7 discusses the presence of VOCs in soil and the three common methods of analysis in use for these substances. This chapter also reviews the various sampling and sub-sampling options for VOCs. Chapter 8 covers non-halogenated organic compounds including semi-volatile organic compounds (SVOCs). The difficulties of pre-treatment and extraction of a wide range of organics from various soil matrices are discussed. The potential use of screening test group parameters (such as extractable organic halogens (EOX) and adsorbable organohalogens (AOX)) is highlighted. The concept of bioavailability is also discussed. Advice on appropriate analysis techniques for the various groups of substances covered in this chapter is also given.

Chapter 9 addresses the issue of leaching tests. The vast majority of contaminated land analysis relates to *total analysis*. This does not give any indication of the availability of the tested substances to biota, groundwater or humans. Leaching tests attempt to simulate the leaching that takes place in the natural environment. It is important to appreciate that leaching methods (unlike total analysis methods) are empirical in that the result is defined by the method, and small changes in the methodology (e.g. changes in leaching temperature, degree of agitation during the leaching tests, method of filtration of the leachate, etc.) can have a significant effect upon the results. The actual leachate used in the test also has a very significant effect upon the results.

The final chapter follows on from leaching tests and looks at the potential uses of toxicity tests in contaminated land analysis and covers ecological assessment and toxicity screening. This holistic approach is a rapidly expanding area that can give useful information on the overall potential toxicity, suggesting the presence of toxic substances, which may not have been analysed for. It can be used for detecting potential contamination missed by conventional targeted and screening chemical analysis and also to confirm that bioremediation processes have been effective in reducing toxicity.

Microbiological analysis is not covered in this book. Microbiological parameters are seldom requested (or needed) for risk assessment, although they do have a role to play when invoking monitored natural attenuation. Occasionally, requests for testing for anthrax spores are received. This is usually in relation to sites that have been associated with certain animal wastes and/or hides. A competent laboratory capable of working with Class 3 (highly pathogenic) organisms is required to carry out this analysis.

1.2 Definition of analysis

The simplest definition of the analysis process is the reduction of uncertainty in the concentration of a substance. Taking cadmium as an example, if we look at the data from a very large database of cadmium analyses of potentially contaminated soils in the UK, over 95% of the results are below 50 mg/kg. Thus, without any further analysis the cadmium concentration, with 95% confidence, of any given random soil sample submitted to a laboratory can be stated to have a cadmium level of 25 ± 25 mg/kg (i.e. up to 50 mg/kg). After analysis this uncertainty is considerably reduced (e.g. 5.1 ± 0.5 mg/kg). In general, the lower the detection limit and the uncertainty required the more costly the analysis. It is also important to appreciate that the sampling errors or natural spatial variability are often more significant sources of uncertainty than the chemical analysis errors in contaminated land site investigations. In general, if the analytical error is less than one-third of the sampling error and small scale spatial variability, further reduction of the analytical error is not considered important (Kratochvil *et al.* 1986).

1.3 Quality issues in contaminated land analysis

1.3.1 Background

Quality has a number of definitions:

1. Fitness for intended purpose.
2. Conformance to specification.
3. Degree of excellence.
4. Meeting agreed customer requirements.
5. The totality of features and characteristics of an entity that bear on its ability to satisfy stated and implied needs (ISO 8402).

In the authors' opinion, the first definition is the most concise and self-explanatory. All analysis should be fit for the intended purpose. A laboratory should be able to determine the concentration of the specified parameter with sufficient accuracy (lack of bias) and precision (repeatability) so that the result can be satisfactorily applied in any relevant risk assessment or remediation verification. The laboratory should also be able to detect the specified parameter at concentrations where there is no significant risk from that parameter on the site in question to the receptor(s) of concern.

1.3.2   ISO 17025

All soil analysis laboratories should be accredited to EN ISO/IEC 17025 (2000), the international standard developed to ensure that analytical calibration and testing conforms to defined professional standards across the world. This standard specifies the general requirements for the competence of laboratories to carry out tests, including sampling. It covers testing performed using standard methods, non-standard methods and laboratory-developed methods. Laboratories complying with ISO 17025 will operate a quality system for their testing and calibration activities that also meets the requirements of ISO 9001 (2001).

The following requirements are covered in the ISO 17025 system.

Management requirements
Organisation
Quality system
Document control
Review of requests, tenders and contracts
Sub-contracting of tests and calibrations
Purchasing services and supplies
Service to the client
Complaints
Control of non-conforming testing and/or calibration work
Corrective action
Preventive action
Control of records
Internal audits
Management reviews

Technical requirements
General
Personnel
Accommodation and environmental conditions
Test and calibration methods and method validation
Equipment
Measurement traceability

Sampling
Handling of test and calibration items
Assuring the quality of test and calibration results
Reporting the results

The basic definition of a quality system is to:

1. Write what is to be done.
2. Do what is written.
3. Write what was done.

All aspects of the analysis must be documented and traceable in the laboratory records including all raw data used to derive the analysis result; the method used; the analyst(s) who carried out the analysis; the dates on which the sample is received, analysed and reported; the instrument used to carry out the relevant analysis; the service history of the instrument; the training record of the relevant analysts to demonstrate their ability to carry out the specified analysis. All records must be kept/stored for a specified time after the completion of analysis. Typically this is for three to six years.

1.3.3   The six valid analytical measurement (VAM) principles

Analytical measurement is of major economic significance with over 10 billion Euros spent annually on chemical analysis in the UK. However, evidence suggests that many analytical measurements are not fit for purpose and that poor quality data represent a major cost and risk to business and society. Consequently, the valid analytical measurement (VAM) programme (Sargent & MacKay 1995; Burgess 2000) was set up in the UK by the Laboratory of the Government Chemist.

There are six succinct basic principles:

1. Analytical measurements should be made to satisfy an agreed requirement.
2. Analytical measurements should be made using methods and equipment which have been tested to ensure that they are fit for the purpose.
3. Staff making analytical measurements should be both qualified and competent to undertake the task.
4. There should be a regular independent assessment of the technical performance of a laboratory.
5. Analytical measurements made in one location should be consistent with those elsewhere.
6. Organisations making analytical measurements should have well-defined quality control and quality assurance procedures.

The six VAM principles are said to provide a framework to enable organisations to deliver reliable results first time, every time, and achieve bottom line improvements through increased operational efficiency and reduction in risk.

1.3.4   MCERTS

The Environment Agency of England and Wales requires operators of regulated processes to deliver monitoring results that are valid, reliable, accurate and appropriate. To this end, the Environment Agency has established its Monitoring Certification Scheme (MCERTS) to improve the quality of monitoring data. MCERTS defines performance standards for continuous emissions monitoring systems (CEMS) and for continuous ambient air quality monitoring systems (CAMS). It also includes a product certification scheme to European Standard EN 45011 and ISO/IEC 17025. This is an expanding scheme to cover all areas of regulatory monitoring including water quality monitoring, manual stack emissions monitoring, data recording and operators' monitoring arrangements. It also covers the competency of personnel for manual monitoring. Ultimately MCERTS will provide a comprehensive framework for industry for choosing suppliers of monitoring systems and services, which meet the Agency's performance standards.

Part IIA of the Environmental Protection Act 1990 (implemented in England and Scotland in 2000 and in Wales in 2001) and the Pollution Prevention and Control (England and Wales) Regulations 2000 require testing to establish the concentration of particular contaminants in soil. As part of the further development of MCERTS, the Environment Agency is in the process of extending the scheme to the chemical testing of contaminants in soil by establishing a register of qualifying laboratories. Initially, MCERTS will not cover sampling or screening tests. Qualification would be by third party accreditation to Agency documented performance requirements based on the European and international standard ISO/IEC 17025:2000. A draft MCERTS laboratory performance standard (Environment Agency 2002b) has been developed to provide an explanation and interpretation of the generally stated requirements of EN ISO/ IEC 17025:2000 for the chemical testing of contaminants in soils. This should be finalised sometime in 2003.

Regulators will rely upon the data produced by laboratories to make key regulatory decisions. It becomes increasingly important, therefore, that appropriate information is provided and that data produced are reliable, and uncertainties associated with their production are explicitly stated. High quality data will be needed to support key regulatory decisions.

Methods for the chemical testing of soils are developed and published by many organisations, but there is, generally, no guidance on the levels of performance these methods should achieve, and whether their performance is appropriate for use, especially for the assessment of potentially contaminated land. Tables 1.2–1.4 give the draft performance characteristics required for metals and metalloids, inorganics and organics (Environment Agency 2002b). This allows laboratories to use any validated analysis method as long as it meets the stated minimum performance characteristics.

Table 1.2 Performance characteristics (metals and organometallics) (Environment Agency (2002b) Performance standard for laboratories undertaking chemical testing of soil. Feb 2003, Version 2)

Parameter[1]	Precision[2]	Bias[3]	Minimum CLEA SGV[4] (mg/kg)
Antimony	7.5	15	–
Arsenic	7.5	15	20
Barium	5	10	–
Beryllium	5	10	–
Boron (water soluble)	5	10	–
Cadmium	5	10	1
Cobalt	5	10	–
Copper	5	10	–
Chromium	5	10	130
Iron	5	10	–
Lead	5	10	450
Manganese	5	10	–
Mercury	5	10	8
Molybdenum	5	10	–
Nickel	5	10	50
Organolead compounds	15	30	–
Organotin compounds	15	30	–
Selenium	7.5	15	35
Thallium	5	10	–
Vanadium	5	10	–
Zinc	5	10	–

[1]Whilst no limit of detection has been specified, it shall be fit for purpose, especially, for example when compared to *soil guideline values* or critical levels of interest.

[2]Precision expressed in percentage terms. (Maximum allowed.)

[3]Bias (or recovery if appropriate) expressed in percentage terms. (Maximum allowed.)

[4]CLEA SGV = Contaminated land exposure assessment soil guideline values. The values given are the minimum values and relate to use of land as *Allotments* or *Residential with plant uptake*. See page 22. Relates to Cr(VI) and inorganic mercury.

An MCERTS register of accredited laboratories will be maintained by the Environment Agency and made available on the web. Accreditation would not be generic. If a laboratory were to seek accreditation for one particular parameter, it would then not be able to analyse any other parameters under MCERTS unless it also gains accreditation for the other relevant specific parameters. The main benefit of MCERTS is establishing a level-playing field in the form of a MCERTS performance standard. This will send a clear message that the production of defensible data for the chemical testing of soils is a crucial component of the regulatory requirements. It will provide assurance to all stakeholders including contractors, regulators, laboratories and the public on the reliability of analytical data generated under MCERTS. This would appear to be a very sensible approach to ensuring that fit-for-purpose analysis data is

Table 1.3 Performance characteristics (inorganics) (Environment Agency (2002b) Performance standard for laboratories undertaking chemical testing of soil. Feb 2003, Version 2)

Parameter[1]	Precision[2]	Bias[3]
Easily liberated cyanide	15	30
Complex cyanide	15	30
Sulfide	15	30
Sulfate	10	20
Sulfur	10	20
Thiocyanate	15	30

[1]Whilst no limit of detection has been specified, it shall be fit for purpose, especially, for example when compared to *soil guideline values* or critical levels of interest. At the time of writing, no contaminated land exposure assessment soil guideline values (CLEA SGV) for these inorganics had been published.

[2]Precision expressed in percentage terms. (Maximum allowed.)

[3]Bias (or recovery if appropriate) expressed in percentage terms. (Maximum allowed.)

provided. It is to be hoped that the scheme will be quickly extended to cover sampling and in situ testing of potentially contaminated land sites.

1.4 Sampling point frequency considerations

The British Standard on the investigation of potentially contaminated sites BS 10175:2001 states that greater confidence in the site assessment can be achieved by increasing the number of samples taken and analysed, as significant differences in the sample composition over small areas in the site can occur. It goes on to state that the errors associated with sampling in site investigations are generally greater than those associated with the analysis. Therefore, it can be more informative to analyse a greater number of samples using simple methodology fit for the purpose rather than analyse a smaller number of samples using a more accurate and costly method.

In order to obtain some perspective of the task involved, BS 10175:2001 (p. 21) suggests a typical sampling grid density at 16–25/hectare for main investigations. Using the lower sampling frequency and the assumptions below,

Number of sampling points per hectare $= 16$
Sampling depth $= 3\,m$
Soil density $= 2.5\,tonnes/m^3$
Number of depths sampled per sampling point $= 5$ (e.g. 0.15, 0.5, 1, 2 and 3 m)

The following can be derived:
Mass of soil assessed per hectare down to $3\,m = 75\,000$ tonnes
Mass of the site soil associated with each submitted sample $= 938$ tonnes $[(75\,000/(5 \times 16))$ samples per hectare]

Mass of typical submitted a litre soil sample $= 2.5\,kg$
Fraction of the site submitted to lab $= 0.0003\%$ $(3 \times 10^{-4}\%)$

Typical mass of samples used for a given analysis $= 1$–$20\,g$
Fraction of site actually analysed $= 2 \times 10^{-6}$–$10^{-7}\%$
(i.e. as little as one ten millionth of a percent or 1 mg per tonne of soil!)

Many site investigations are carried out at a sample point frequency significantly less than this. It can be seen that if the ground is heterogeneous, the

Table 1.4 Performance characteristics (organics) (Environment Agency (2002b) Performance standard for laboratories undertaking chemical testing of soil. Feb 2003, Version 2)

Parameter[1]	Precision[2]	Bias[3]
Benzene	15	30
Benzo[a]pyrene	15	30
Chlorobenzenes	15	30
Chloromethane	15	30
Chlorophenols	15	30
Chlorotoluenes	15	30
Dichloroethanes	15	30
1,2-dichloroethene	15	30
Dichloromethane	15	30
Dioxins	15	30
Ethylbenzene	15	30
Furans	15	30
Hexachloro-1,3-butadiene	15	30
Hydrocarbons	15	30
Nitroaromatics	15	30
Pentachlorophenol (PCP)	15	30
Phenols	15	30
Phthalate esters	15	30
Polyaromatic hydrocarbons	15	30
Polychlorinated biphenyls	15	30
Tetrachloroethanes	15	30
Tetrachloroethene	15	30
Tetrachloromethane (carbon tetrachloride)	15	30
Toluene	15	30
Trichloroethanes	15	30
Trichloroethene	15	30
Trichloromethane (chloroform)	15	30
Vinyl chloride	15	30
Xylenes	15	30

[1]Whilst no limit of detection has been specified, it shall be fit for purpose, especially, for example when compared to *soil guideline values* or critical levels of interest. At the time of writing no contaminated land exposure assessment soil guideline values (CLEA SGV) for organics had been published. See Editors' note on page 176.
[2]Precision expressed in percentage terms. (Maximum allowed.)
[3]Bias (or recovery if appropriate) expressed in percentage terms. (Maximum allowed.)

result of the analysis may be highly questionable. The number of sample point locations should represent a balance between the degree of confidence required and the total cost of the proposed sampling and analysis, given the likely heterogeneity of ground conditions and the processes dispersing the contaminants in the ground. Variograms may be used to demonstrate the presence of spatial correlation of the sample spacing used and to quantify small-scale variability (Nathanail 1997).

1.5   Sample pre-treatment issues

The sample preparation protocol is an integral part of the analysis result. This important criterion is often overlooked by users of the analysis results. For example, there are three common soil preparation protocols:

1. *Protocol 1*: All results relate to the air-dried part of the soil that will pass through a 2 mm-mesh sieve after gentle manual crushing to break up larger soil particles. All stones larger than 2 mm are assumed to be inert and rejected prior to analysis of the ground (less than 200 μm) portion. Results relate to the sub-sample analysed.
2. *Protocol 2*: As protocol 1, but using a 10 mm sieve. All stones and other components larger than 10 mm are assumed to be inert and rejected prior to analysis of the ground (less than 200 μm) portion. Results reported in relation to the sieved sub-sample analysed.
3. *Protocol 3*: As protocol 2 except all results reported in relation to *as received* sample, not the sub-sample analysed.

If we assume that

1. A given sample has 70% of *stones* >2 mm.
2. A given sample has 30% of *stones* >10 mm.
3. The CLEA cadmium soil guideline value (SGV) for residential land use (without plant uptake) of 30 mg/kg is the criterion being used to assess the samples.
4. The true cadmium content is 20 mg/kg in the total air-dried sample.
5. The stones contain a negligible amount of cadmium.

Then, following the three protocols (assuming accurate analysis) will yield the following results:

Protocol	Cadmium concn mg/kg	Exceedance of SGV
1	66.6	Yes
2	28.6	No (borderline)
3	20	No

Given that the SGV is driven by ingestion of soil and indoor dust, protocol 1 best reflects the likely exposure to cadmium in the potentially contaminated soil. Thus, it is very important to ensure that the correct sample pre-treatment protocol is adopted and that the end-user of this data understands this protocol and how the analysis is reported. Not all laboratories clearly specify the protocol that they use. It is also essential that the fraction of the sample discarded prior to sub-sampling for analysis is recorded in the analysis certificate. Also, details of any unusual artefacts or large debris that were removed prior to the analysis pre-treatment stages should be included in the analysis certificate.

It is important that soil samples are taken in appropriate sample bottles/ containers. Samples for volatile organic compounds should be taken in sealed glass vials, filled to the top to minimise any headspace, to ensure that volatile components cannot escape. Also special precautions have to be taken for other labile and unstable parameters. Prior to commencing a site investigation, risk assessors should liaise with their analytical laboratory to establish what sample containers should be used.

Conservative, non-labile parameters such as toxic metals, chloride, water soluble boron, elemental sulfur are analysed on the air-dried sample ($30 \pm 5°C$) after crushing and grinding, whilst labile and more volatile substances are analysed on the *as received* sample and the results converted to an air-dried basis after air drying a representative separate aliquot of the sample. The sampling for these labile and more volatile substances in the *as received* sample is more prone to error (i.e. lower precision) than sampling an air-dried, crushed and ground sample but ensures that the analyses are fit for purpose with respect to bias.

In the UK it is normal practice to quote all results relative to the air-dried sample at $30 \pm 5°C$, whereas elsewhere in Europe it is more usual to quote the results against a separate sub-sample dried at $105 \pm 5°C$. For many samples this does not make a big difference (typically less than 10%) but for a few samples significantly larger differences can be observed between the two drying techniques. The laboratory sampling errors associated with dried sieved and ground samples are significantly less than those with *as received* samples containing significant amounts of heterogeneous matter. For example, attempting to sub-sample 2–10 g from an oily, tarry, stony *as received* soil samples is notoriously difficult. For these and other very heterogeneous *as received* samples, it is recommended that replicate sub-samples should be taken and analysed to improve the precision of the analysis. Laboratories should advise clients, when in their view, this is required. The client can then decide whether to accept the increase in analysis cost or risk receiving poorer quality data.

It is generally agreed that an air-dried ($30 \pm 5°C$) sample is satisfactory for toxic metals and other non-labile parameters such as elemental sulfur. Loss or degradation of many organic parameters (e.g. phenols, VOCs, TPH, etc.) and other labile parameters, such as easily liberatable cyanide, can occur on drying (30±5°C) followed by grinding. Also, the leaching properties of the sample can change during these processes. For certain parameters that are borderline (e.g. naphthalene in the PAH suite, mercury and DRO) laboratories should ensure that their air drying and grinding procedures do not result in significant losses of these determinands.

1.6 Analysis method requirements

All analytical methods should be:

1. Fit for purpose.
2. Robust and reliable.
3. Not prone to significant matrix interference effects.
4. Adequately validated for all relevant matrixes.
5. Ideally automatable (to make them cost effective).

The biggest problem in contaminated land analysis is the wide range of matrices encountered (e.g. clay, peat, limestone, sandstone, steel slag, waste materials, demolition debris). It can be difficult to ensure that unacceptable biases do not arise for a small percentage of samples because of the wide range of matrices that can be encountered. Various quality assurance/quality control protocols can be used to attempt to minimise these effects (see Chapter 2). The use of appropriate certified reference soils or lower cost well-characterised in-house soil reference materials for quality control purposes is strongly recommended. Figures 1.1 and 1.2 attempt to depict the concepts of precision and bias. A repeat analysis of a highly biased yet precise method result will

Good precision, negligible bias

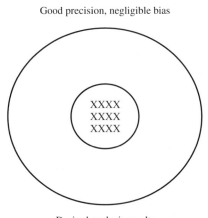

Desired analysis results

Figure 1.1 Precision and bias of results concept.

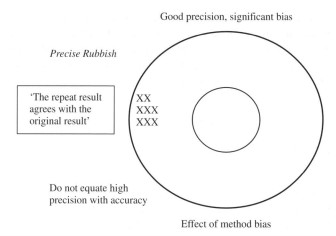

Figure 1.2 Precision and bias of results concept.

give a similar, but inaccurate result. Another key area is the estimation of uncertainty of measurement of a given method (see Section 1.7). Increasing use of low bias, low precision, low cost methods can result in increased certainty in the conceptual site model through the greater sample density arising from the low cost of each analysis.

1.7 Measurement uncertainty

Measurement uncertainty is an estimate attached to a measurement which characterises the range of values within which the true value is asserted to lie. Every measurement has an uncertainty associated with it, resulting from the variability arising in the various stages of sampling and analysis and from imperfect knowledge of factors affecting the result. For measurements to be of practical value, it is necessary to have some knowledge of this measurement uncertainty. A statement of the uncertainty associated with a result conveys to the customer the quality of the result and allows an evaluation of fitness for purpose to be made.

A statement of uncertainty is a quantitative estimate of the limits within which the true value of a measurand (such as an analyte concentration) is expected to lie. Uncertainty is expressed as a standard deviation or a calculated multiple (typically twice) of the standard deviation as an uncertainty range. In obtaining or estimating the uncertainty related to a particular method and analyte, it is essential to ensure that the estimate explicitly considers all the possible sources of uncertainty. Repeatability and/or reproducibility, for example, are not generally acceptable as estimates of the overall uncertainty, since neither takes into account any systematic (bias) effects inherent in a method. (In general, this is often because only a few matrices are used for validation exercises prior to analysis. It is assumed that corrections are made for known systematic (bias) errors, but with contaminated land analysis with a very wide range of potential matrices this can be very difficult.)

All testing laboratories should apply and report the outcome of procedures for estimating the uncertainty of measurement. For soil analysis, because of the wide variation in matrices encountered, it may not be possible to devise a rigorous, metrologically and statistically valid calculation of measurement uncertainty. The laboratory should attempt to identify all the significant components of uncertainty and make a reasonable estimation. They should ensure that the form of reporting the result does not give a wrong impression of the uncertainty, e.g. a soil zinc result of 1455.55 mg/kg would be better reported as 1460 mg/kg, or a soil PCB result of 0.1576 mg/kg would be better reported as 0.16 mg/kg.

In general, results should be reported to such a level that rounding errors do not occur when results are used in statistical calculations. Reasonable estimation of uncertainty can be based on knowledge of the performance of the method and can make use of (11 day) method validation data (Cheeseman & Wilson 1989) and/or proficiency scheme data. This assumes in both cases that an adequate number of soil matrix types are tested. It can be very instructive to compare the uncertainties produced by these two methods. Figure 1.3 attempts

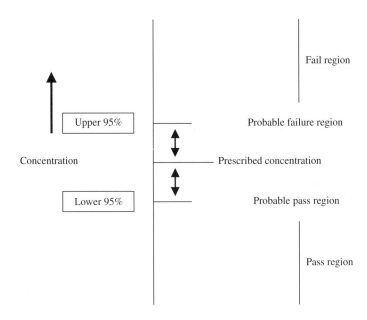

Figure 1.3 Importance of result confidence limits.

to show the significance of uncertainties with respect to meeting specific statutory values or assessment criteria.

1.8 Proficiency schemes

It is important that laboratories participate in proficiency schemes (ISO/IEC 1997). Proficiency schemes are run by specialist organisations that circulate homogenised soil samples to participating laboratories on a regular basis (typically 5–12 times a year). The laboratories then have to analyse the submitted sample(s) within a given time period (typically one month) and submit their results to the scheme. The scheme organiser then sends each laboratory a summary of all the results received but only identifies the laboratory to which the results are sent (i.e. the results of all other laboratories are anonymous). From the results, the scheme organiser will indicate what is considered to be an acceptable range of the results. This will clearly highlight unfit-for-purpose results. Thus laboratories can assess their performance against other laboratories. Some schemes, in addition to soil samples, also distribute solution method calibration standards, and digests or extracts of the distributed soil. The results from these additional samples should help the laboratory to determine the cause of a failure of the soil sample, for example the failure caused by incorrect preparation of standards, poor extraction of the analyte from the soil matrix, or sample matrix interference effects.

It is essential that all proficiency scheme samples are treated as *normal samples* and are analysed along with the daily routine samples. One of the key purposes of a proficiency scheme is that laboratories thoroughly audit all failures and attempt to determine the cause of any failures. A laboratory should then implement suitable protocols to prevent further failures from the same cause(s). The laboratory should regularly monitor its cumulative proficiency scheme results to ensure that there are no consistent failures for any given determinands and make the results of its monitoring available to its risk assessor clients.

It is important for laboratory clients to examine their results closely as soon as possible (within three days) after receipt of the analysis data. They should then query the laboratory on any questionable results. This will allow the laboratory to check for calculation and transcription errors and if necessary to re-analyse the relevant sample extract/digest or carry out a full re-analysis of the soil sample. It is frustrating (but quite common) for laboratories to receive queries long after (6–18 months) the samples were analysed when it is no longer practical for re-analysis. All that can be done is to check for transcription and calculation errors.

Another concern is that some laboratory clients are requesting that a league table of proficiency results by laboratory be drawn up and published. The

danger with this approach is that some laboratories will then treat the proficiency samples as *special samples* and carry out replicate analysis using their
most experienced analysts. The proficiency results no longer reflect the *typical
performance* of the laboratory, but the *best performance possible* of the laboratory. This information is not what the client or the regulator requires. The
fairest way to test laboratories would be to use truly blind proficiency samples,
where laboratories are not aware that the analysed samples are proficiency
samples. To date, it has not been possible to do this for a significant time
period because of the cost and logistics involved. Indeed, what is needed is not
a league table but confidence that a laboratory is above a preset performance
threshold.

It is felt that the frequency of proficiency samples (5–12 distributions/year)
is not really adequate as many large laboratories analyse 200–1000 soil samples
per day. Most proficiency schemes are either accredited to or working towards
accreditation by the end of 2003. This accreditation will assure the user that
the scheme meets adequate quality technical standards. Finally, it should be
pointed out that the proficiency scheme soil samples are supplied as finely
divided ground-dried sample (to ensure all supplied samples are equivalent
and homogeneous). Thus, it cannot always be assumed that because a laboratory
obtains fit-for-purpose results on proficiency samples it will also obtain fit-for-
purpose results for all the range of the samples that it receives, especially for
parameters that are analysed on an *as received* (wet weight) basis. Many
analytes will be more readily extracted from the supplied finely ground proficiency sample than many *as received* real sample matrices. This is one reason
why laboratories should validate their methods using a wide range of typical
samples.

1.9   New areas of analysis

With the new risk-based approach to assessing contaminated land sites, there
are a number of new areas of analysis that are required by risk assessors.

1.9.1   *Organic carbon*

The fraction of organic carbon is a soil and site specific parameter that affects
the fate and transport of organic substances. The greater the fraction of organic
carbon the higher the site specific assessment criterion for volatile substances.
There are two main methods for determining this: by combustion of an aliquot
of the soil after removing inorganic carbon (carbonate) using acid and measuring the evolved carbon dioxide; (ISO 10694:1995) and oxidation of soil organic
matter by gently boiling with a solution of potassium dichromate, sulfuric acid
and orthophosphoric acid. The excess dichromate is determined by titrating
with ferrous sulfate solution (MAFF 1985).

1.9.2   Partition coefficients (Kd values)

These describe the rate of contaminant transport relative to that of ground-water. The Kd parameter is used to estimate the potential for the adsorption of dissolved contaminants in contact with soil. Kd is usually defined as the ratio of the contaminant concentration associated with the solid to the concentration of the contaminant in the surrounding aqueous solution when the system has reached equilibrium. A simple laboratory batch method can be used to determine this. A known volume and concentration of aqueous contaminant (adsorbate) is added to a known mass of soil in a sealed container. The sample is shaken for a pre-determined equilibration time at a constant temperature. The soil is separated from the adsorbate and the remaining concentration of contaminant in the adsorbate is determined. Control samples of the contaminant with no soil present are also set up and analysed to allow for any degradation of the aqueous contaminant with time. There are also a number of much more elaborate methods for carrying out this measurement. The reader is referred to the USEPA website http://www.epa.gov/radiation/cleanup/partition.htm entitled *Understanding variation in Partition Coefficient, Kd, Values* for further information on this topic. It is important to appreciate that these methods like all leaching-based tests are empirical and the results can be strongly influenced by the extraction protocol used (see page 216).

1.9.3 Bioavailability and bioaccessibility

Bioavailability is defined by Kelley *et al.* (2002) as 'the extent to which a chemical can be absorbed by a living organism'. The Environment Agency (2002a) defines the term as 'the fraction of the chemical that can be absorbed by the body through the gastrointestinal system, the pulmonary system and the skin'. In practice, it is almost impossible to estimate or measure the bioavailable portion of a contaminant. Bioaccessibility represents a halfway house that can be estimated under laboratory conditions. Bioaccessibility is the fraction of a chemical that is dissolved from a soil sample using *in vitro* (*test tube*) test methods that simulate gastrointestinal conditions (Kelley *et al.* 2002). The Environment Agency (2002a) describes this as 'the fraction of a substance that is available for absorption by an organism.' Bioaccessibility is used as a cautious estimator of relative bioavailability.

Bioaccessibility testing of metals using the physiologically based extraction test (PBET) or the simplified bioaccessibility extraction test (SBET) has gained popularity in recent years in the UK (Nathanail & McCaffrey 2002). The aim of the tests is to simulate the extraction of metals into solution in the juices in the stomach, upper and lower intestine (PBET) and in the stomach (SBET). The empirical test results should only be applied to the ingestion of soil or dust pathways. The tests have been calibrated for lead, and to a lesser degree for arsenic. Risk assessors should be satisfied that the health criteria value or the generic or site specific assessment criterion do not already make an allowance for bioaccessibility. For example, the generic CLEA[1] soil guideline value for lead uses an empirically derived linear model relating lead intake to observed blood lead concentrations in adults and children. The tests should only be applied to specific pathways and not to all contaminants.

The protocols for carrying out the PBET (Ruby *et al.* 1996) or IVG (*in vitro* gastrointestinal test) (Rodriguez *et al.* 1999) are extremely complex and involve a large number of reagents, pH changes, solution agitation, maintaining anaerobic conditions, etc. They are empirical methods where the result is likely to vary with small changes in test conditions (e.g. method of agitation, pH used, temperature, sample pre-treatment protocol etc.) In addition, the PBET for lead and arsenic uses an oven dried (50°C sample) sieved to <250 μm, with a 0.4 g aliquot of test material used. The perceived uncertainty limits of these tests are considered to be high. In the authors' opinion, there is a need for suitable certified reference materials to be made available and for proficiency schemes to encompass these tests. Users of PBET and SBET results should use such results with considerable caution.

This is illustrated in a paper by Oomen *et al.* (2002). This paper described a multi-laboratory comparison and evaluation of five *in vitro* digestion models. The experimental design and the results of a round robin evaluation of three soils, each contaminated with arsenic, cadmium and lead were presented and discussed. A wide range of bioaccessibility values were found for the three soils:

	Soil 1 (%)	Soil 2 (%)	Soil 3 (%)
For arsenic	6–95	1–19	10–59
For cadmium	7–92	5–92	6–99
For lead	4–91	1–56	3–90

Details of the five *in vitro* methods used [SBET (BGS); DIN method (RUB); *in vitro* digestion model (RIVM); SHIME method (LabMET/Vito); and TIM method (TNO)] are given in the paper.

Note

1. For further information on publications relevant to the Environment Agency of England and Wales contaminated land research (CLR) R&D programme, see page 22.

References

Barr, D., Bardos, R.P. & Nathanail, C.P. (2002) Non biological methods for assessment and remediation of contaminated sites: case studies. Funders Report. CIRIA, London.

BS 5930 (1999) *Code of Practice for Site Investigations*, British Standards Institution (London).

BS 10175 (2001) *Investigation of Potentially Contaminated Sites*. Code of practice, British Standards Institution (London).

Burgess, C. (2000) Valid analytical methods and procedures, Royal Society of Chemistry. ISBN 0–85404–482–5.

Cheeseman, R.V. & Wilson, A.L. (Revised by Gardner, M.J.) (1989) A manual on analytical quality control for the water industry. NS 30. Water Research Centre.

DoE (1994) Contaminated Land Research Report 4, *Sampling Strategies for Contaminated Land*, DoE (London).

EN ISO/IEC 17025 (2000) General requirements for the competence of testing and calibration laboratories.

Environment Agency (2000a) *Technical Aspects of Site Investigation Volume I (of II) Overview* R&D Technical Report P5-065/TR. ISBN 1 85705 544 6.

Environment Agency (2000b) *Technical Aspects of Site Investigation Volume II (of II) Text Supplements* R&D Technical Report P5-065/TR. ISBN 1 85705 545 4.

Environment Agency (2002a) Contaminated Land Exposure Assessment (CLEA) home page. www.environment-agency.gov.uk (search on "CLEA").

Environment Agency (2002b) Performance standard for laboratories undertaking chemical testing of soil. Feb 2003, Version 2. This will be revised (see website http://www.environment-agency.gov.uk/business/mcerts/).

Ferguson, C., Darmendrail, D., Freier, K., Jensen, B.K., Jensen, J., Kasamas, H., Urzelai, A. & Vegter, J. (eds) (1998) *Risk Assessment for Contaminated Sites in Europe Volume 1 Scientific Basis*. LQM Press, Nottingham. ISBN 0 9533090 0 2.

Finnamore, J., Barr, D., Weeks, J. & Nathanail, C.P. (2002) *Biological methods of assessment and remediation of contaminated land*. CIRIA Report C575, CIRIA London.

Heath, J.S., Koblis, K. & Sager, S.L. (1993) Review of chemical, physical and toxicologic properties of components of total petroleum hydrocarbons. *J. Soil Contamination*, **2**(1), 1–25.

ISO 10694 (1995) Soil quality – Chemical methods – Determination of organic and total carbon after dry combustion (elementary analysis) (BS 7755-3.8:1995).

ISO 9001 (2000) Quality management systems – Requirements.

ISO TR 13530 (1997) – Water quality – A guide to analytical quality control for water analysis.

ISO/DIS 10381 (2002) Soil quality – Sampling –
  Part 1: Guidance on the design of sampling programmes.
  Part 2: Guidance on sampling techniques.
  Part 3: Guidance on safety.
  Part 4: Guidance on the procedure for investigation of soil contamination of urban and industrial sites.
  Part 5: Guidance on investigation of soil contamination of urban and industrial sites.
  Part 6: Guidance on the collection, handling and storage of soil for the assessment of aerobic processes in the laboratory.

ISO/IEC Guide 43-1 (1997) Proficiency testing by interlaboratory comparisons – Part 1 Development and operation of proficiency testing schemes.

Kelley, M.E., Brauning, S.E., Schoof, R.A. & Ruby, M.V. (2002) Assessing oral bioavailability of metals in soil. Battelle Press, Columbus. ISBN 157477123X.

Kratochvil, B., Goewie, C.E. & Taylor, J.K. (1986) Sampling theory for environmental analysis, *Trends in Analytical Chemistry*, **5**(10), 253–257.

MAFF (1985) The analysis of agricultural materials, Reference book 427, Method 56, Organic Matter in Soils, HMSO, London. ISBN 0 11 242762 6.

Nathanail, C.P. (1997) Expert knowledge to select model variogram parameters for geostatistical interpolation of sparse data sets. In: Yong, R.N. & Thomas, H.R. (eds), *Geoenvironmental Engineering; Contaminated Ground: Fate of Pollutants and Remediation*, Thomas Telford, pp. 240–247.

Nathanail, C.P. & McCaffrey, C. (2002) Use of oral bioavailability in assessment of risks to human health from contaminated land. *Environment 2002*. Available from www.environment-2002.com.

Oomen, A.G., Hack, A., Minekus, M., Zeijdner, E., Cornelis, C., Schoeters, G., Verstraete, W., Van De Wiele, T., Wragg, J., Rompelberg, C.J.M., Sips, A.J.A.M. and Van Wijnen, J.H. (2002) Comparison of five *in vitro* digestion methods to study the bioaccessibility of soil contaminants. *Environ. Sci. Technol.*, **36**, 3326–3334.

Persoone, G., Janssen, C. & De Coen, W. (2000) New microbiotests for routine toxicity screening and biomonitoring, Kluwer Academic/Plenum Publishers, New York 2000. ISBN 0–306–46406–3.

Ramsey, M.H. (1998) Sampling as a source of measurement uncertainty: techniques for quantification and comparison with analytical sources. *J. Analyt. At. Spectrometry*, **13**(2), 97–104.

Rodriguez, R.R., Basta, N.T., Casteel, S.W. & Pace, L.W. (1999) An in vitro gastrointestinal method to estimate bioavailable arsenic in contaminated soils and solid media. *Environ. Sci. Technol.*, **33**, 642–649.

Ruby, M.V., Davis, A., Schoof, R., Eberle, S. & Sellstone, C.M. (1996) Estimation of lead and arsenic bioavailability using a physiologically based extraction test. *Environ. Sci. Technol.*, **30**, 422–430.

Sargent, M. & MacKay, G. (1995) Guidelines for achieving quality in trace analysis, Royal Society of Chemistry, 1995. ISBN 0–85404–402–7.

Publications relevant to the CLR R&D programme

Available from http://www.defra.gov.uk/environment/landliability/index.htm

Main reports

CLR 7 Assessment of risks to human health from land contamination: an overview of the development of guideline values and related research, March 2002.

CLR 8 Priority contaminants report, March 2002.

CLR 9 Contaminants in soil: collation of toxicological data and intake values for humans, March 2002.

CLR 10 The contaminated land exposure assessment model (CLEA): technical basis and algorithms, March 2002.

TOX reports (see also Appendix 3)

TOX 1 Arsenic, March 2002.
TOX 2 Benzo(a)pyrene, April 2002.
TOX 3 Cadmium, March 2002.
TOX 4 Chromium, March 2002.
TOX 5 Inorganic Cyanide, March 2002.
TOX 6 Lead, March 2002.
TOX 7 Mercury, March 2002.
TOX 8 Nickel, March 2002.
TOX 10 Selenium, March 2002.

Soil Guideline Value (SGV) reports (see also Appendix 3)

SGV 1 Arsenic, March 2002.
SGV 3 Cadmium, March 2002.
SGV 4 Chromium, March 2002.
SGV 5 Inorganic Mercury, March 2002.
SGV 7 Nickel, March 2002.
SGV 9 Selenium, March 2002.
SGV 10 Lead, March 2002.

2 The requirements of the analytical method

David Westwood

This chapter sets out to provide advice and guidance to analysts and others on the selection of appropriate methodology, on why and how methods should be fully validated, and to provide stimulus to those who consider what actions may be necessary to ensure the correct interpretation of data. There is little justification for carrying out any analysis unless there is confidence in the results generated, the results are fit for purpose, and meaningful decisions can be made on the strength of the confidence shown in the results. Within this chapter, less emphasis is placed on the *how and what* of methodology than on the *why* of correct methodological practice.

2.1 Need for fully documented and properly validated methods

When analytical requirements for soil (including contaminated land) analysis are considered, two main factors need to be taken into account before analysis can begin. Firstly, a method must be shown to be suitable for use with the matrix under investigation. This is not an easy task and can raise practical problems that need to be addressed. Secondly, before a method is used routinely to generate results within a laboratory, it must be fully validated in the laboratory where it is used. In addition to these considerations and practical difficulties, the method must be suitable for the particular determinand, analyte or parameter being analysed, and this includes the concentration range and level of interest. Thus, the laboratory should demonstrate that the method is suitable and fit for purpose.

Without fully documented and properly validated methods, analysts, or others who use the data provided by analysts, will not have the confidence to ensure that the results that are generated are fit for the purpose for which they were originally intended. Methods should be fully documented, and should clearly and unambiguously describe the determinand being determined as well as the concentration range of the determinand for which the method is applicable, and the matrix, or matrices, to which the method can be applied. In addition, clear details of the procedures to be followed should be documented in such a manner that the text cannot be open to mis-interpretation and that the analyst can precisely carry out that which needs to be done. Without all this information, the scope of the method cannot be clearly defined and the method will often be misused, for example, in the inappropriate determination of parameters (e.g. by arbitrarily adding another metal to a metal suite) or matrices (e.g. validating a method with a sandy soil and inappropriately applying it to a peat-based soil), or in the determination of parameters at inappropriate concentrations.

Within the environmental sector, there are two major reasons why inappropriate methods are so often used. Firstly, the analysis of soils (including contaminated land) is carried out in a fiercely competitive and price-driven market. Secondly, there is no clear message or direction given from regulators, or contractors and consultants who procure analytical services. Regulators are faced with huge problems when dealing with issues that are essentially of a site-specific nature. These problems include the specification or prescription (for all sites) of the requirement of a single suite of determinations. In many instances, the specified suite of determinations may not be appropriate for specific individual sites. In addition, it would appear that most contractors and consultants would rather be told what specific analyses need to be carried out than take responsibility for their own requirements.

In a price-driven competitive market, laboratories are caught between the need to secure business and the need to provide good quality analytical services that are fit for purpose. Hence, quality issues assume less priority and importance in order for laboratories to become more efficient and cost-effective. Consequently, it is often the case that a single method will be validated for a specific determinand in a very narrow range of soil matrices, often for a limited concentration range, but used routinely for all circumstances. To ensure laboratory efficiencies and satisfy imposed contractual restraints, this method is then applied and used in situations where similar related determinands or matrices are analysed. Unfortunately, it often happens that the method is used to analyse not so similar determinands or matrices. Hence, the method becomes used in inappropriate circumstances and the results generated are no longer fit for purpose.

A classic example of this commonly adopted approach is the determination of a metallic trace element in a defined solid environmental matrix. This determination may involve an optimised digestion stage pertinent to a particular element and a specific matrix. Because this method may be the only validated method available, the digestion procedure is then used to determine other suites of metals in other matrices. Whether the actual procedures used are appropriate to these other metals in the suite or indeed the matrix is often not fully appreciated, or even considered, and results are generated in the hope that they will be satisfactory for the purpose. Without some form of verification, this assumption may be totally erroneous and the results generated may not be fit for purpose.

Without a regulatory driver to reduce cost considerations and impose minimum quality requirements, the situation with respect to contaminated land analysis is unlikely to change quickly. In addition, without contractors and consultants clearly specifying exactly what analyses are required for their particular needs and circumstances, the widespread provision of quality analytical services will be delayed for a considerable time. As contract analysis is, generally, awarded on cost considerations, until the philosophy of contractual analytical procurement changes, improvement in quality services will be limited. Notwithstanding this, it is slowly being recognised that quality issues are important, and play a major role in focussing attention on the confidence that stakeholders, regulators, process operators and the public demand, when assurance is sought on the reliability of test data.

2.2 Current regimes

The majority of contaminated land analyses depend upon site-specific requirements. Only when the history of the use of individual sites has been determined can the analytical requirements be seriously considered. Since no two sites are exactly the same, and because all have individual aspects that need individual considerations, it is unlikely that analytical requirements will be identical. In addition, the matrix of samples taken from a particular site may be different for different samples, as well as for different sites. This variety in the nature of matrices and analytical requirements makes the imposition of either prescribed *standardised suites of analyses* or the specification of methodology somewhat meaningless; as in many cases, these would not be appropriate.

Current laboratory practices revolve around the use of accredited methods, and indeed, these practices need to be supported and complimented. However, the use of an accredited method in itself is not a guarantee of the generation of good quality data. Methods, whether accredited or not, need to be used in correct circumstances, especially in their applicability to specific or defined matrices.

A long-established approach gaining more popularity is the recognition that performance-based methodology has practical benefits that lead to improvements in the quality of data. The specification of minimum performance standards eliminates the need to specify defined methodology, but prescribes the minimum performance requirements that all methods should satisfy. Such a scheme based on this approach has recently been introduced by the Environment Agency under its Monitoring Certification Scheme (MCERTS) which is related to the chemical testing of soil, and further details of this scheme can be found on the Agency's internet web page (Environment Agency 2003). The scheme, based on laboratory accreditation (BS EN ISO/IEC 17025 2000), contains additional requirements to those contained in the international standard to ensure the competency of laboratories and the reliability of generated data.

2.3 How to validate international, national and individual laboratory methods

Method validation is not, as is often assumed, carried out to ensure whether the performance characteristics of a method are fit for purpose, but rather to establish the potential capability of a method (in terms of accuracy, precision etc.) in the actual laboratory where the validation is undertaken. When the performance characteristics have been established, a judgement can then be made on whether they are fit for an intended purpose. The capability established in one laboratory is not transferable to another and it does not mean that the same capability can be achieved in another laboratory. Only when full inter-laboratory performance testing of a method is undertaken can the overall capability of the method be established.

A great deal of guidance and information has already been produced on the validation of methods, most of it, unfortunately for environmental analysts, directed towards water and related industries, and it is not the intention of this chapter to duplicate this literature with respect to soil analysis (BS EN ISO/IEC 17025 2000; Burgess 2000; Cheeseman & Wilson 1989; Codex 2002; Eurachem Working Group 1995; FAO/IAEA 1997; Holcombe 1998; ISO 5725). Whilst much of this advice on water methodology can be applied to soil (including contaminated land) analysis, there are certain differences which become apparent when the procedures are applied to soil. Water is often regarded as a simple homogeneous substance, and whether the subtle differences in nature resulting from the source of the water are taken into account is debatable. For example, the differences noted between upland (humus-containing) waters, ground waters and lowland (river) waters are not often appreciated, and a method used for a particular type of water is often (quite wrongly in certain cases) used for other types of waters (e.g. pesticide analysis).

With soil, the situation is further complicated and exacerbated because of the varied nature of the matrix and by the fact that in most circumstances the definition of soil is not, and indeed cannot be, clearly and unambiguously stated. Depending upon the natural composition of the sample and its level of contamination, each sample removed from an individual site can be regarded as being different from other samples, either from the same site but from different locations within the site, or from different sites. This makes the process of validation extremely difficult when applying these procedures to soil matrices. Concerns are then raised over the number or type of matrices that require validated methodology, and often these issues cannot be addressed when considered on cost grounds alone. In addition, whilst validation principles basically remain the same, irrespective of matrix, the application to different matrices can cause enormous practical problems. For example, spiking experiments involving soils are far more complex compared with aqueous samples, where for the vast majority of non-saline waters, samples comprise over 99.5% of the sample mass. For soils, the situation is much more complicated.

If carried out correctly, the process of validation enables performance characteristics, such as bias, recovery, precision and limit of detection, to be estimated with a high degree of confidence. Whether the performance characteristics determined are deemed acceptable, or not, is another matter but the actual values determined can only be assessed when the process has been completed. Hence, full method validation should only be attempted when the procedures to be validated have been shown, at least initially, to provide results that are deemed acceptable. If the performance characteristics are not acceptable, either from a regulatory or fit-for-purpose point of view then alternative procedures should be sought and included in further method development or the method abandoned for one more suitable; either way, further validation is required. Hence, prior to routine analysis being undertaken, some degree of method assessment should be carried out to demonstrate that the method might be applied to a particular matrix or determinand of interest. Only when method development or optimisation is complete and this demonstrates that acceptable results can be generated should full validation be contemplated.

Only when fully validated methods are used can confidence be generated in the results that are determined. Validation procedures involve a series of procedures to enable performance characteristics to be calculated. These procedures may involve repeated analysis, undertaken over a period of days or weeks, of certified reference materials, and/or spiking experiments involving blank and inert or similar matrices to those being analysed but containing negligible amounts of determinands of interest. Having established the full capability of the method for the matrix and determinand of interest, analysts should demonstrate that the performance of the method is maintained throughout its use within the laboratory. This can be established using appropriate internal quality control analysis. In addition, greater confidence can be demonstrated by participation in relevant external quality control or proficiency testing schemes. More important than the validation process itself, however, is the act of demonstrating that the use of the method is restricted to the appropriate matrix and determinand as defined in its scope.

The level of confidence gained from validated methodology is dictated by the amount of relevant information provided via the validation process. Thus, the more the information, the greater the confidence in the data. In turn, the more confidence in the analysis, the better the foundation for proper and meaningful decisions to be made using the data. It is often queried how much information is required in order to provide greater confidence and indeed this is a difficult question to answer; for example, whether a minimum number of degrees of freedom should be specified in the generation of data for the determination of performance characteristics. Depending on the specific values obtained, it may be that from a statistical point of view, there is little difference whether ten or twenty degrees of freedom are used to generate the performance characteristics. Whilst a higher number of degrees of freedom may demonstrate some improvement in the quality of the data produced, there will be huge cost considerations involved and any benefits to the improvement in the quality of data might be far out-weighed by over-burdensome costs.

As already stated, prior to the start of validation, it is essential that the procedures to be undertaken be clearly and unambiguously documented. Also, when the process of validation commences, it is equally essential that the procedures that are documented be strictly adhered to. Minor changes to the procedures should not be introduced, as these could affect, adversely or otherwise, the resulting performance. The only way by which changes to the performance of a method can be assessed would be to validate a method and establish its performance, then to introduce changes to the procedures and then re-validate the method to establish whether a change in the performance had occurred.

Another major problem associated with validation procedures is that of establishing bias and recovery values. For water analysis, these two values are separate and distinct. However, for soil analysis they may be regarded as being synonymous. Bias can be estimated using certified reference materials. Whether the certified reference material is aqueous or solid is immaterial, provided it is appropriate to the matrix and concentration level under investigation. However, at present, there are very few certified reference materials covering the environmental sector with respect to all sample matrices and determinands, and determinands of different concentration levels. In addition to cost considerations, the lack of appropriate certified reference materials is often blamed as the other main contributing factor to why more validated methods are not used.

The estimation of recovery, carried out using spiked samples with known amounts of determinands, is often the only alternative to estimating bias when certified reference materials are not available. Notwithstanding this, spiking experiments should be used with caution, since the added determinand may not be in the same form as that contained in the matrix. In addition, the time between the addition of the determinand and its extraction needs to take into account, and allow for, potential occurrence of matrix–determinand interactions. This is not always fully appreciated, and in fairness, is very difficult to accommodate. Questions like 'how long should a spiked addition be left before extraction?' and 'what form should the added determinand comprise?' are extremely difficult to address.

Notwithstanding whether the method is an in-house method that has been provided/developed, a national method, or indeed an international method, validation of the method should be undertaken in the laboratory where the method is to be used. This is to ensure that the laboratory can demonstrate its capability in carrying out the method. Ideally, all national and international methods should have been inter-laboratory performance tested to establish their full capability or performance characteristics. These methods should be clearly and unambiguously written and the procedures strictly adhered to when performance testing is undertaken. However, in reality, this may not be the case for either situation.

2.4 Quality control – quality assurance

Whenever a laboratory has validated a method that is to be used routinely, that laboratory should also demonstrate that it is capable of maintaining the stated performance of the method. This can be achieved in several ways and includes the analysis of certified reference materials and spiking experiments involving the analysis of samples before and after addition of known amounts of the determinand being analysed. Whilst the analysis of blank and standard solutions enables the detection or determination stage of the method to be assessed, the whole method can only be assessed by analysing similar matrices to those under investigation. Alternatively, the assessment may involve spiking experiments where samples are analysed before and after the addition of known amounts of the determinand being determined. This involves using the entire method including procedures for preparation and pre-treatment, extraction and concentration, and the detection or determination stages. The results of the internal quality control samples are then plotted on control charts and warning and action limits established.

Warning and action limits are often based around standard deviations of the mean of the relevant control samples, usually two and three times the standard deviation respectively, and these limits are plotted on the control charts. If these limits are breached, an investigation should be undertaken to ascertain the cause and, if appropriate, remedial action carried out to ensure that future violations are eliminated. Rules can be set up to facilitate the interpretation of a potential breach, for example, two consecutive results lying outside warning limits but inside action limits, or the appearance of a specified number of descending or ascending results. It is recognised that a small number of breaches may be explained statistically. For example, with a 95% level of confidence, one breach outside the warning limit but inside the action limit may be explained by the fact that one result in twenty may occur by chance and in reality no remedial action need be taken. The establishment of these so-called rules is for the benefit of analysts, and once established, should be used in a consistent manner to facilitate confidence in the data generated. Ideally, the data used to generate the warning and action limits should be generated prior to the analysis of real samples. In practice, results from samples and quality control results are generated together, and the warning and action limits, initially, are based on estimated values. Periodically, the control charts should be re-plotted using up-to-date quality control data to generate warning and action limits.

2.5 Prescribed method versus minimum performance characteristics approach

In order to demonstrate comparability in contaminated land analysis, there is a view expressed that methods should be specified or prescribed, thus ensuring that all analysts use exactly the same methods. However, without some form of *policing system* involving an audit or control process, even the prescription of methodology does not guarantee strict adherence to specified methodology. In the case of a regulator prescribing, for example, a statutory method, the regulator needs to be assured that for every occasion the method is prescribed, and that it is appropriate and fit for purpose. Even if this were to be the case, there would be occasions, however rare, when the use of a prescribed method was incorrect in its application. This could then severely restrict or even hinder the enforcement of particular legislation. In the context of soil (including contaminated land) analysis, to undertake this approach the regulator would need to be confident that when a method was prescribed, it was appropriate to all the matrices and parameters for which the method was applied. This would be extremely difficult, if not impossible, to achieve.

In addition, unless frequently updated, the specification or prescription of methodology hinders technological developments until newer methods or technologies can be incorporated and taken on board. Since legislative processes are extremely slow to progress, until this could happen, analysts would only be allowed to use out-dated methods based on old technologies.

An alternative approach to the prescription of methodology is to allow analysts the freedom to use whatever method the analyst feels is suitable, provided the method is fully validated, satisfies minimum performance criteria and is shown to be fit for purpose. This approach leaves the analyst the choice of which method to use, dependent only upon the needs and requirements of the analysis pertaining to the circumstances of the analysis. At any time, using this approach, laboratory staff can adapt any method it is considered appropriate to do so to ensure that the method used is pertinent to the sample matrix being analysed and the determinand and concentration level of interest being determined. The responsibility for using the most appropriate method therefore resides with laboratory staff, in consultation with the consultant or contractor requiring the analysis.

2.6   Proficiency testing for contaminated soil analysis

Precision within a laboratory can be tightly controlled by the operation of internal quality control schemes. This can be achieved by undertaking replicated analysis of samples and/or possibly by analysing replicate samples. The analysis of certified reference materials offers a means whereby bias can be estimated. Another means of estimating a laboratory's performance is via participation in inter-laboratory proficiency-testing schemes. Originally, external quality control or inter-laboratory proficiency-testing schemes had been made available to assist laboratories to identify those areas where remedial action was considered necessary, in order to improve the quality of data provided. More recently, however, the results from inter-laboratory proficiency-testing schemes have been widely used to assess the performance of a laboratory and to show how well that laboratory performs. An assumption is then made that if a laboratory performs consistently badly in a scheme, then its day-to-day routine capability should be questioned.

Ideally, all analyses, whether consisting of real samples or internal or external quality control samples, should be carried out under exactly the same conditions, using the same staff, reagents, equipment, etc. Unfortunately, in order to demonstrate good performance in external quality control or inter-laboratory proficiency-testing schemes, some laboratories allow their more able analysts to analyse the samples supplied by the proficiency-testing scheme organisers. Alternatively, if it is recognised that a sample from such a scheme is being analysed, greater care is observed in the detail of the analysis. Whichever approach is adopted, a misleading assumption on the routine capability of the laboratory is provided. Only when all staff analyse the routine samples and proficiency-testing scheme samples, will a true indication of the laboratory performance be forthcoming. Occasionally, a totally *blind approach* is adopted by scheme organisers or individual organisations. This approach is based on the commercial distribution of samples where the laboratories concerned are unaware that the samples being analysed are part of a proficiency-testing scheme, and treat the samples in the normal manner undertaken for routine commercial samples.

Generally, a proficiency-testing scheme organiser will distribute, to participating laboratories, samples consisting of standard solutions, prepared extracts of matrices, and homogenised matrices to be extracted. All solutions are then determined by the method of choice within the individual laboratory. A comparison is then made on all the results submitted by the participating laboratories. In this way, an assessment of bias can be made on whether the laboratory can successfully analyse standard solutions or, using a consensus mean of all results, how laboratories compare in their analyses of prepared extracts and homogenised matrices. The assessment of bias is then carried out using a statistical treatment to calculate *z-scores*. Values of *z-scores* are then interpreted as reflecting satisfactory, questionable or unsatisfactory performance.

One of the major disadvantages of using a consensus mean value instead of the true value is that it can never be stated with absolute certainty whether a result which appears to be outside the range of acceptability is in fact closer to the true value than the consensus mean. However unlikely, it may be that a majority of laboratories have generated poor results and a small number of laboratories have generated good results but it is the minority of laboratories that are assessed as poor performers. This situation could be avoided if certified reference materials were distributed. However, there are few reference materials that closely match or resemble all relevant matrices and determinands of interest, including concentration levels.

2.7   Reference materials

Certified reference materials offer an ideal means of estimating bias within a laboratory. In addition, other reference or in-house materials, if assessed or evaluated against certified reference materials, offer an alternative means of assessing a laboratory's capability and establishing a means whereby the analysis is judged to be within acceptable limits. Unfortunately, there are insufficient certified reference materials available to match the wide-ranging variety of matrices and determinands analysed routinely within the environmental sector. Depending on how the reference materials are used, the capability of individual analysts or laboratories can be determined.

Usually, certified reference materials are reported with a certified value of a determinand, together with narrow tolerance limits within which the result of the analysis should lie. The certified value may be obtained as a mean of values reported by a variety of methods, or it may be a method-based value determined using a specific method. For example, this may involve extracting the determinand with a known solvent composition.

Certified reference materials should be used in the validation process to establish the capability of the method and then as part of on-going quality control procedures to ensure that the capability of the method is maintained in routine operation. Alternatively, reference or in-house materials (which are often less expensive) can be used in on-going quality control procedures, provided the materials have been evaluated against the certified reference material.

References

BS EN ISO/IEC 17025 (2000) General requirements for the competence of testing and calibration laboratories.

Burgess, C. (2000) Valid analytical methods and procedures. The Royal Society of Chemistry. ISBN 0–85404–482–5.

Cheeseman, R.V. & Wilson, A.L. (Revised by Gardner, M.J.) (1989) A manual on analytical quality control for the water industry, NS30, Water Research Centre, Henley Road, Medmenham, Marlow, Bucks, SL7 2HD. ISBN 0–902156–85–3.

Codex (2002) Codex committee on methods of analysis and sampling. In-house method validation (document CX/MAS 01/9). See also Codex committee on methods of analysis and sampling. Consideration of IUPAC guidelines for the in-house (single laboratory) validation of methods of analysis (document CX/MAS 02/10).

Environment Agency (2003) http://www.environment-agency.gov.uk/business/mcerts.

Eurachem Working Group (1995) Quantifying uncertainty in analytical measurement, LGC, Queens Road, Teddington, Middlesex, TW11 0LY. ISBN 0–948926–08–2.

FAO/IAEA (1997) Validation of analytical methods for food control. A report of a joint FAO/IAEA expert consultation. December 1997, Vienna.

Holcombe, D. (1998) The fitness for purpose of analytical methods. A laboratory guide to method validation and related topics, LGC, Queens Road, Teddington, Middlesex, TW11 0LY. Eurachem Working Group.

ISO 5725. Accuracy (trueness and precision of measurement methods and results) Parts 1–6.

Part 1:1994 – General principles and definitions.

Part 2:1994 – Basic method for the determination of the repeatability and reproducibility of a standard measurement method.

Part 3:1994 – Intermediate measures of the precision of a standard measurement method.

Part 4:1994 – Basic methods for the determination of the trueness of a standard measurement method.

Part 5:1998 – Alternative methods for the determination of the precision of a standard measurement method.

Part 6:1994 – Use in practice of accuracy values.

3 Initial sample preparation

Mark Allen

3.1 Introduction

Modern analytical methods can provide a rapid and highly sensitive examination of contaminated soils, sediments, landfill, etc., provided that the samples presented for analysis are homogeneous and in an appropriate physical form.

Initial sample preparation is the means by which samples are transformed from a *raw* or *as received* state into a form suitable for chemical analysis. Because this involves direct manipulation of sample powders by handling, sub-sampling or comminution, the quality of the initial sample preparation techniques employed strongly influences the quality of the analytical service provided to customers. Indeed, inappropriate application of sample preparation methods may entirely invalidate the determination of some analytical parameters. For example, inappropriate drying or comminution may cause a significant loss of volatile or labile parameters. Samples collected from contaminated sites vary widely in their physical form and may include vegetation, animal remains, plastics, soil, metallic scrap and building rubble. Sample preparation methods need to be planned to account for these variations so that a representative, homogenous sub-sample can be prepared as efficiently as possible.

Initial sample preparation can be differentiated broadly into four main processes.

(a) *Reception and initial handling.* A visual examination of the samples, followed by labelling and recording the sample's, size, type, texture, general condition and the suitability of packaging are often undertaken during the registration of samples into the laboratory quality assurance and health and safety systems.

(b) *Pre-preparation.* Normally designed to terminate chemical and biological processes occurring in the samples and to *prime* the material for any subsequent preparation processes. With the exception of samples containing labile or volatile parameters, this may be achieved by freezing, air drying, oven drying, freeze drying, autoclaving, screening or particle size distribution analysis (sieve testing). Removal and recording of any material unsuitable for preparation, the extraction of pore waters and sampling the *as received* samples for labile or highly mobile parameters e.g. VOCs, PCBs, PAHs are often carried out at this stage.

(c) *Comminution*. Crushing and milling processes are used to improve sample homogeneity by mixing and reducing the maximum particle size of the samples to allow representative sub-sampling. It should be noted that milling is not appropriate if important analytical parameters are affected by changes in the surface area of the samples (e.g. BTEX, interstitial gases, etc.).

(d) *Finishing*. The preparation of finely milled analysis samples, pressed powder pellets, etc. suitable for loading directly into analytical instrumentation, or appropriate for analytical pre-treatment (e.g. leaching or fusion into glass beads). The quality of *finished* samples is carefully verified prior to their transfer from the preparation laboratory to analysts.

The high productivity achieved by modern analytical instrumentation means that sample preparation is typically slower and considerably more labour intensive than chemical analysis. As a result there is often a considerable incentive to prepare samples as rapidly as possible, because the costs of analytical *down time* are usually very high. Unfortunately, there is a potential for error (e.g. an unwanted change in the composition of the sample) at each stage of the initial sample preparation process and great care must be taken to ensure that the need for high productivity does not overshadow the need for quality in the preparation methods used.

Because initial sample preparation is essentially practical in nature and because the processes utilised are often considered to be well understood, the potential for serious error during sample preparation is frequently underestimated. The provision of a sufficient supply of representative sub-samples for analysis irrespective of sample type involves applying a *fit for purpose* method for each individual sample prepared. The following chapter describes some of the most widely used initial sample preparation methods and proposes practical solutions to problems encountered in the preparation laboratory.

3.2   An overview of initial sample preparation

Before initial sample preparation commences, it is essential that the customer and analyst confirm the purpose and nature of the sample preparation processes needed to ensure that analytical and risk assessment objectives are achieved.

When a detailed site assessment is needed, multi-element trace level analysis is often essential. In this case the samples may be sub-sampled directly from their containers and analysed without further sample preparation to preserve labile and volatile parameters. For more easily retained parameters, a fine powder (e.g. 99% <30 microns) is produced by a series of processes including crushing, sub-sampling, milling and fine milling using non-metallic milling vessels under virtually dust free conditions. Up to half the total preparation time may be spent ensuring the cleanliness of equipment to minimise contamination and cross-contamination during processing. This type of preparation is relatively slow but produces fully representative finely powdered sub-samples with the lowest practicable levels of contamination. Alternatively if a general profile of the site is needed, sub-sampling for volatile/labile parameters is followed by rapid particle size reduction (e.g. down to about 80% <100 microns) using larger scale metallic milling techniques requiring no sub-sampling with the exception of removing a *grab* sample of the milled powder for analysis. This method is much more rapid, less costly and involves less cleaning time.

The flow diagram in Figure 3.1 identifies the processes commonly applied to contaminated land samples as they are prepared for chemical analysis. Generally samples are prepared for only two or three analytical techniques, so it is unlikely that an individual sample would be subjected to all the processes shown.

3.2.1 Receipt of samples by the laboratory

Immediately on receipt of samples by the laboratory, sample details are normally entered on a laboratory information management system (L.I.M.S.) and a quality control system. The processes to which the samples are to be subjected, the condition of samples on receipt and their suitability for preparation are recorded and the appropriate administration is initiated. Even at this early stage it is not uncommon for unexpected charges to be incurred by the customer. Typically these charges relate to handling difficulties caused by the leakage of samples during transport, or by illegible or complex sample labels which inhibit the transfer of sample numbers into the sample administration system. Given that manpower is often the largest component in the total cost of sample preparation, these problems may result in a significant increase in the total cost of sample preparation services. Simple sample codes are a major benefit throughout the analytical process as they greatly reduce the potential for errors caused by mis-reading labels or when transcribing labels from one container to another. Despite the operation of computer-controlled labelling systems, errors in labelling can be significant particularly when multiple sub-samples are generated from a single original sample.

Any preparatory work that may be done whilst collecting samples will reduce the cost of sample reception (e.g. drying, compositing samples, selecting duplicate or replicate samples, etc.).

3.2.2 Pre-preparation

Samples containing labile or volatile parameters (e.g. VOCs, PCBs, PAHs, methane, etc.) are typically subjected to minimal handling to prevent unwanted alteration of the sample before analysis. For this reason, analysis sub-samples are often carefully taken in a pragmatic manner directly from appropriate

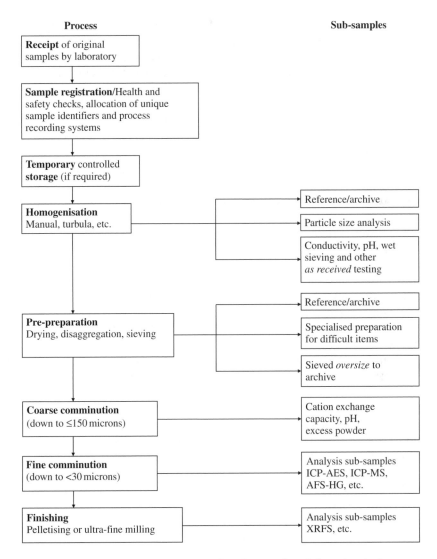

Figure 3.1 Initial sample preparation: flow diagram of regularly used methods.

sample containers with no initial sample preparation (ISO/FDIS 14507 2002; Standing Committee of Analysts 2003).

It may be necessary to retain some of the solid for reference purposes or subsequent further analyses. A sufficiently large sub-sample should be taken to ensure that the portion analysed is as representative of the original sample as possible. If the *as received* samples contain a significant amount of large or heterogeneous debris, replicate sub-samples should be taken and analysed to improve the quality of the data.

The analyst may wish to remove and record particles *naturally* larger than a given particle size. Typically, for soils, particles larger than 10 mm or 2 mm are removed by mechanical or hand sieving and their mass recorded as a proportion of the total dry mass of the sample. Care should be taken to avoid removal of aggregates formed during drying.

Depending on the composition of each individual sample, it may be necessary to remove large pieces of highly resistant material including metal, concrete and plastics at this stage. If these items are considered to be an important part of the sample, as in samples of incinerator feed, they will need to be prepared using specialised crushing, cutting or milling methods, prior to returning them in powdered form to the bulk sample for mixing and further processing.

If the major portion of the sample consists of large pieces of metal, plastic slag etc., this may necessitate changes in analytical procedures, for example, the methods used to prepare samples for analysis by dissolution techniques (e.g. ICP-AES, ICP-MS, AFS-HG, etc.) may need to be modified to prevent saturation of analytical equipment. Good communication between the sample preparation laboratory and analytical departments may save the effort wasted on re-analysing samples that could be expected to give an atypical response during analysis.

For many analytical applications it is essential to comminute the whole sample by one means or another before sub-sampling. This avoids the possibility of serious sub-sampling error caused by removing part of an inhomogeneous sample.

Different laboratories have different routine protocols to overcome this problem but whatever the processes employed, initial sample handling must minimise the loss or chemical alteration of important analytical parameters (e.g. reduction of sulfates, biological breakdown of nitrates, volatilisation of As, Sb, Hg and volatile organic parameters).

Depending on the analytical requirements, samples may be *stabilised* in readiness for further preparation by freezing, freeze drying, air drying or oven drying at low temperature (typically 30–35°C) until the sample weight is approximately constant (i.e. a within 5% variation on repeated weighings).

Freeze drying is most efficient when samples are frozen before drying. The method is useful for fine grained samples including clay-rich soils, sediments and filter cakes; it should be noted that this technique is not suitable for volatile organic parameters. The process involves cooling the samples to around −30°C in vacuum followed by gentle heating to ambient temperature (normally about 25°C). The sublimation of water from the samples tends to shatter the fine aggregates present in most contaminated soils producing a coarsely divided powder. The advantage here is that the sample is not *baked* into a hard briquette. Subsequent disaggregation and milling is therefore much more straightforward and will result in less sample contamination because the samples will cause less wear to milling equipment. Samples of wet soil weighing less than 500 g can normally be freeze dried in a porous container (e.g. paper bag) in 24–36 hours.

For some parameters, such as PCBs, PAHs, mercury and selenium, drying is normally carried out at less than 35°C. Unfortunately, low temperature drying can be prolonged, but assuming the samples are protected from air-borne contamination this may be the best method for preventing loss of these parameters from the samples. Laboratories must carry out validation trials to ensure that no significant loss of these and other relevant parameters occurs during the sample preparation procedure.

Fan-assisted oven drying is the most widely used and productive drying method; generally temperatures of 35°C ± 5°C or 105°C ± 5°C are used. Drying times vary from a few minutes to 4 or 5 days. The quality of this method is reliant on careful calibration and use of the drying ovens. Disadvantages include the need to periodically monitor ovens for hot or cold areas and difficulties in disaggregating or comminuting the sample after drying. Care should be taken to ensure that trays or other containers do not contaminate or absorb any part of the samples.

Plant materials are commonly dried by oven drying, microwave drying or freeze drying, and the friable product is then easily comminuted using high shear milling techniques. Microwave and freeze drying methods have been found to give similar performance when preparing samples for determination of non-volatile organic parameters (e.g. sugar alcohols, organic and amino acids) (Popp *et al.* 1996).

For practical reasons, achieving a dry sample is important. Samples containing more than about 0.5% free water, and/or about 0.1% liquid organic material (e.g. oils, tars, biological residues, etc.) are not amenable to crushing or milling and without appropriate drying, preparation cannot be completed effectively. Samples holding non-aqueous liquids may be super-cooled immediately before comminution or crushed when frozen in freezer mills if appropriate.

3.2.3   Coarse comminution

The main purpose of coarse comminution in sample preparation is to reduce the maximum particle size of samples to allow effective mixing prior to subsampling. The ratio of feed size before reduction to the product size after reduction is known as the reduction ratio, and for the first crushing or milling stage this ratio should be as large as possible (Smith & James 1981). Ratios of 40/1 for coarse crushers and 15/1 for coarse milling equipment are useful minimum specifications for laboratory preparation of coarse contaminated soil samples of up to 2 kg.

The simplest way to reduce the samples to a usable coarse powder is by hand in the tried and trusted mortar and pestle. This is the most sensitive and easily controlled crushing method. If required, samples may be disaggregated using metallic or ceramic equipment without grinding harder particles. The limitations with this method are sample hardness and operator safety. Up to 60 soft or granular samples may be disaggregated by hand in a typical working day; however, productivity falls quickly when sample hardness rises. Hand disaggregation may generate air-borne dust, causing both a risk to the safety of the technician and a potential for carry over of material from one sample to another. Dust control measures must be carefully considered when undertaking sample preparation processes of any kind.

For higher productivity or harder samples, a wide range of semi-automatic crushing and milling equipment may be used to break up the samples. Larger clasts (e.g. stones) or aggregates above 180 mm maximum particle size may need to be crushed using a jaw or roller type crusher. Jaw crushers use a gyrating jaw working in opposition to a fixed jaw in the form of an off-centre V-shape with a *nip point* at the base of the V. The sample particles are gravity fed from a hopper into the crushing chamber where they are crushed until they are small enough to pass through the nip point (Fig. 3.2). Roller crushers use heavy counter-rotating metallic rollers to reduce large, hard or tough particles down to nominally 2 mm maximum particle size.

If highly resistant items (e.g. metal bars, rubbers or plastics) are to be included in the sample, they require preparation by a more appropriate method. Metals may be carefully cut down to <3 mm using a water-cooled saw or metal shear followed by cleaning to remove any contaminating swarf. Milling using

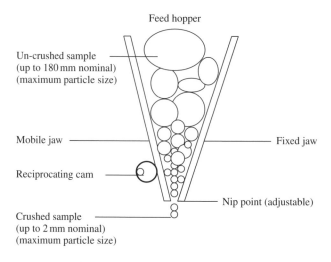

Figure 3.2 Jaw crusher – mode of operation.

a high powered, high-shear knife, hammer or pin mill will then reduce the cuttings to a coarse powder suitable for return to the original sample for homogenisation. Rubbers, plastics, fabrics and other soft, flexible materials may be super-cooled in liquid nitrogen or simply frozen until brittle, before crushing. Again, high-shear milling is generally effective.

Typically dried coarse soil samples of <40 mm maximum particle size are reduced in disk, ring, rod, shatterbox or Tema type high abrasion mills using steel, tungsten carbide, zirconium oxide, hard porcelain, silicon nitride, boron carbide or agate milling vessels (other materials are available). These vessels contain up to three free moving milling elements normally comprising a large ring enclosing a smaller ring, which in turn encloses a solid puck. The sample powder is placed between the milling elements and the vessel is tightly closed. The loaded vessel is then clamped to the mill and oscillated at up to 1000 rpm (at an amplitude of 20–40 mm) for an appropriate period, normally between two and fifteen minutes depending on the type of sample and the density of the milling vessel (Fig. 3.3). After an appropriate milling time, a moderately fine powder (about 80% less than 150 microns) is produced. Up to four samples may be reduced in each milling cycle.

Using a shatterbox or Tema type mill fitted with a chromium-steel vessel, it is possible to reduce four 200 g samples of metallic slag up to 10 mm in size to less than 200 microns in a 10 minute milling time. A higher power, higher volume mixer type mill may reduce a 1500 g sample of this material to a comparable product size in five minutes. Both the methods are efficient but introduce metallic contamination to the sample. Agate milling vessels do not add metallic contamination but milling is much less efficient due to the lower density of agate when compared with other milling materials. An illustration of the relative efficiencies of milling vessels is given in Table 3.1. The particular

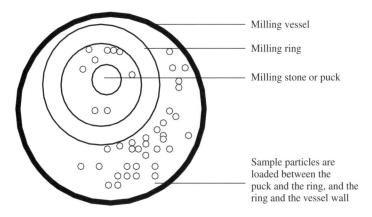

Milling vessel

Milling ring

Milling stone or puck

Sample particles are loaded between the puck and the ring, and the ring and the vessel wall

Figure 3.3 Ring mill (Shatterbox, Tema) – mode of operation.

Table 3.1 Comparison of coarse milling efficiency: Tema milling a 150 g sample of 0.5 mm quartz sand for 10.0 minutes

	Percentage of sample passing test sieve mesh		
Milling vessel	150 μm	125 μm	63 μm
Chromium steel	100	96	44
Colmonoy (Ni alloy)	100	95	40
Tungsten carbide	100	95	46
Silicon nitride	100	99	50
Agate	78	59	18

milling method is selected according to the amount of sub-sample, the productivity required and the acceptable type and amount of contamination.

The high level of abrasion within shatterboxes, Tema or ring mills allows efficient cleaning by milling a charge of clean silica sand. Washing, compressed air blasting and brushing are also regularly used methods for limiting cross-contamination. Sand cleaning and washing are essential for trace level preparation, and compressed air brushing is normally effective for general purposes.

Cone crushers are particularly useful for quickly reducing large volumes of moderately resistant sample (e.g. soil, softer slags and ashes). Comminution occurs when the sample is fed from a hopper into a conical toothed crushing chamber containing a toothed cone. The clearance between the walls of the crushing chamber and the cone is adjustable by raising or lowering the cone. The cone is rotated inside the chamber by a powerful motor and the sample particles are broken between the cone and the walls of the chamber (Fig. 3.4). Reduction ratios of up to 10/1 can be achieved and up to 200 kg of material per day may be ground using a bench-mounted cone grinder fitted with a ceramic cone and chamber.

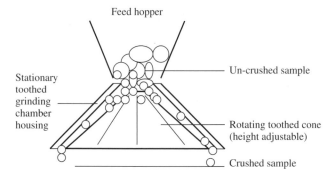

Figure 3.4 Cone grinder – mode of operation.

For samples of engineering soils, clay liners for landfill and other materials where surface area sensitive parameters are important, controlled coarse comminution using slow mill speeds and/or low attrition milling techniques may be used to produce uniformly milled powders suitable for analysis (e.g. determination of cation exchange capacity). The milling time and the amount of sample loaded into the vessel have a major effect on the analytical values obtained, and careful control of the method is needed. For this and other similar applications, the attrition level within the milling vessel can be reduced by removing the puck, creating low attrition conditions or removing a milling ring for a very low attrition milling action. This form of milling tends to liberate individual grains without immediately comminuting them. Low attrition milling may also be used for removing mineralised coatings from hard clasts (stones), or hardened surfaces from soil aggregates; in some cases merely turning or rolling the samples in a roller blender or mixing in a turbula (mixing device in which the sample container follows an end-over-end or figure of eight motion) may produce the desired separation with minimal size reduction. The very large increase in the surface area of the sample caused by its reduction to a fine powder may facilitate the collection of interstitial gases from the sample.

By using modified milling equipment it is possible to collect the gases liberated during comminution of a known amount of solid (Fig. 3.5). Routine

Figure 3.5 Modified milling vessel for liberation of interstitial gases (left) and sample collection vessel (right).

methods for the semi-quantitative estimation of methane in coal chippings and drill cuttings utilising modified Tema milling vessels have been developed. The gas is liberated by a short milling period (up to one minute), which reduces coal chippings to less than 60 microns. The gas is collected after a timed delay to allow the gas to desorb to equilibrium. The short milling period allows full liberation but limits heat generation within the mill vessel (Creedy 1986).

Some milling vessels are specifically designed to be super-cooled in liquid nitrogen to allow difficult samples including hair, rubber or bone to be milled whilst in a brittle condition. Various manufacturers now offer milling vessels, modified to allow size reduction under a controlled atmosphere to prevent redox reactions during milling.

3.2.4   Sieving for preparation purposes

An inspection of published methods suggests that with the exception of samples collected for labile parameters, the majority of samples collected from contaminated sites are screened at either 2.0 mm or 10.0 mm to remove *un-representative items*. Dry screening the disaggregated dried material at 2.0 mm is perhaps the most widely employed method. Particles passing through a 2.0 mm mesh can be nominally described as being within the sand, silt or clay fractions of the sample. An advantage of screening at this size is that an estimation of the relative proportions of these fractions (either visually or by undertaking a particle size analysis) allows a description of the soil texture. Sample texture is an important property that allows a preliminary estimation of the permeability and cation retention capacity of the sample (White 1987). Clearly, the mesh size chosen for screening will have a fundamental effect on the composition of the samples and caution should be taken when drawing comparisons between data derived from samples screened at differing mesh sizes. The method of screening should always be recorded in the final laboratory report. The proportion of any un-sievable material (e.g. metal pieces, robust plastics etc.) should also be recorded separately, or included as part of the oversize fraction if appropriate.

3.2.5   Homogenisation

The difficulties associated with maintaining sample homogeneity when comminuting and sub-sampling coarse soils or other innately inhomogeneous materials are considerable. Hand mixing wet *as received* samples can be done using a large scoop or spatula. Another quick and convenient method suitable for dry powders is to pass each sample through a riffle splitter, re-combining the powder after each pass. Four or five passes are usually sufficient to thoroughly mix the sample. Manual mixing of large samples (>10 kg) by moving the material from one position to another on a clean plastic sheet or smooth concrete floor allows a thorough visual assessment of the overall nature of the sample. A decision may then be made regarding the size to which the samples are to be reduced either by cone quartering or by another suitable method.

Alternatively, mechanised methods for homogenising samples are extremely varied and include Ribbon mixers, V-blenders and Turbula mixers for samples up to 2000 kg. For smaller samples, roller blenders, bench-mounted conical mixers or large food blenders may be suitable. In general, any vessel large enough to contain the entire sample with at least one third of the internal volume available for movement of the powder is suitable. The vessel should be designed to prevent fractionation of particles according to their density, size or shape by means of internal paddles or fins; random shaking of the vessel may also be effective. Practical experience suggests that some homogenising systems, such as small cone blenders, may induce particle separation in some samples.

Finely milled powders (i.e. <30 micron max. particle size) should not suffer significant segregation, although good practice suggests that thorough mixing is advisable after prolonged storage or transport of powders, however fine they are. The only way to verify that a sample is homogenous may be to assess the data generated from a series of sub-samples taken at random from the bulk sample. If a lesser degree of certainty is acceptable, a coloured marker may be added to the sample. The evenness of colouration may then serve to denote the level of homogeneity.

3.2.6   Sub-sampling

Regardless of the analytical requirements, sub-sampling is best avoided by collection of duplicate samples on site if possible. The preparation of each duplicate sample may then be tailored to specific analytical parameters. Often, sub-sampling cannot be avoided and it is necessary to take the largest possible aliquot from a well mixed (preferably crushed or milled) sample. If necessary, the sub-sample may be formed by combining a number of smaller samples for added safety.

Unfortunately, preparation processes (particularly primary crushing) may increase inhomogeneity in the sample. The different components in individual sample may be separated according to their level of resistance to comminution. Relatively high density brittle components may be crushed first and lower density soft components may be crushed last. This effect creates an inhomogeneous powder in which platey and lighter components tend to *float*, and heavy malleable components tend to *sink* within the sample powder. It is therefore important that crushed or coarsely milled material is always well mixed before sub-sampling is carried out. Not surprisingly, unless the sample is inherently homogenous as with some filter cakes and industrial powders, sub-sampling can be a major source of error in sample preparation.

Sub-sampling can be avoided by using large steel or tungsten carbide ring mills, shatterboxes or disk mills which allow up to 2 kg of sample to be milled to 100% less than 150 microns in a single milling cycle. A small *grab* sample of milled and mixed powder may then be removed directly for analysis.

Cone-quartering is a rapid and widely used sub-sampling method, particularly well suited for large samples and it is arguably the most accurate system for sub-sampling (Mutter 1997). The sample is mixed and the powder slowly poured from its container on to a clean surface in the form of an inverted cone. The cone is then carefully split into quarters, using a spade or spatula and alternate quarters are removed. The cone is then re-formed and the process is repeated until a suitable amount of sample forms the cone. A similar method involves forming the powder into a thin layer, which is then separated into four quadrants. Alternate quadrants are removed and the process progresses as above (ISO 11464 1994).

Riffle splitting is also a widely used but less applicable sub-sampling method, its main limitation being the size of the riffle equipment available. To ensure efficient sub-sampling, the splitter should have grid apertures of size at least twice the maximum particle size of the sample. The sample is poured evenly through a riffle grid consisting of a series of chutes which open alternately into one of the two riffle boxes placed beneath the grid. Each time the sample passes through the grid it is sub-sampled in half and the process is repeated until a suitable amount of sample is deposited in one of the riffle boxes. Care must be taken when riffle splitting samples of differing composition. These splitters have many corners which may retain powder; this may then be transferred from one sample to another if the riffle is not correctly brushed out or washed and thoroughly dried, ideally in a low-temperature oven.

Rotary sub-sampling provides the option of generating multiple sub-samples in a single process and is best suited for large, finely divided samples (below 1 mm maximum particle size) or inherently homogenous materials. Rotary splitters vary in design but generally consist of a vibrating feed hopper and a series of delivery shutes. In some designs, the splitter can be adjusted to provide up to 30 sub-samples of nominally equal mass (usually within 0.5% of the target sub-sample mass).

Random sub-sampling or *dipping* is often used for sampling stockpiles and other large volumes, or in the preparation laboratory for sampling bulk unmilled powders. Usually ten to fifteen small aliquots are selected, each aliquot consisting of a few grams of powder. Incremental reduction is similar to dipping but the sample is formed into a thin layer and divided into 20 to 40 square or rectangular portions, and a sub-sample is taken from each portion of the sample and homogenised (Smith and James 1981).

Regardless of the type of sub-sampling method used, some loss of fines may be expected, though, with care dust losses and separation effects should be less likely during coning and quartering when compared with other methods. This is because the sub-sample can be taken without vibrating the powder in a feed hopper (rotary sampling) or pouring the entire sample through a series of chutes (riffling).

3.2.7 Fine comminution

Many analytical techniques now require very small sample sizes, for example, Inductively Coupled Plasma-Atomic Emission Spectroscopy (ICP-AES) and ICP-Mass Spectrometry (ICP-MS) methods may utilise only 0.10–0.5 g of powder which must of course be representative of the original sample. Unless the collected sample is small and finely divided, representivity at this small sample size can only be achieved by fine milling the sub-samples of the coarse powder.

Again, a large number of techniques are available to produce fine powders: shatterboxes and Tema mills may be used, but lengthy milling times (up to 45 minutes) may be needed to reduce powders to 100% <70 microns. Ball milling, centrifugal ball milling and micronising are regularly used techniques but many others are available. Ball milling allows relatively high productivity compared to shatterboxes or Tema mills because up to eight samples may be milled in a single cycle. The amount of sample milled and the milling time tend to be the main factors influencing the maximum particle size of the product. The sample is placed in a milling vessel, containing a number of milling balls (normally one to twenty-five) and the sealed vessel is then clamped into the mill. The mill rotates (or vibrates) the vessel at up to 400 rpm and the sample is reduced by high-energy collisions and abrasion between the milling balls and the walls of the vessel (Fig. 3.6). Typically 80–100 ml milling vessels are used to reduce up to 35 g of dry sample to about 95% <40 microns in eight minutes (metallic milling vessels) or 100% <40 microns in 30 minutes (agate milling vessels). Up to eighteen milling cycles may be completed per day by one technician.

Ball milling is a regularly used method for fractionation of organic matter in soils. The sample is initially dispersed in water using ultrasonic dispersion and then ball milled until a fine slurry is produced (Schmidt et al. 1997). To achieve a smaller particle size it may be appropriate to use grinding aids to increase the level of attrition within the milling vessel (powdered quartz) or to reduce adhesion of the sample to the milling elements (ethylene glycol) (Broton 1996). If appropriate, ball milling the sample as an aqueous slurry at approximately 50% solids can be effective in reducing the maximum particle size to less than five microns, although in agate vessels the milling time may extend to 60 minutes for this level of reduction. Ball milling in a solvent is not advised, as even at low mill speeds some heat generation may occur as the sample is reduced.

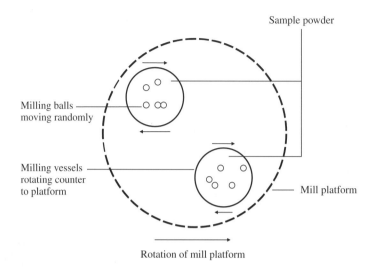

Figure 3.6 Ball mill – mode of operation.

Sub-micron milling is not widely used for contaminated soils; however, semi-quantitative determination of asbestos may necessitate milling to sub-micron level. Micronising mills and other ultra-high energy centrifugal, oscillating ball mills or air jet mills are often used for this purpose.

Preparation for X-ray diffraction studies requires micronising with minimal (preferably zero) introduction of stress to the mineral structures. For soil samples, pre-treatment to remove organic matter, soluble salts or mineral cements may be required (Buhrke *et al.* 1998). Productivity is typically lower than that of other milling procedures because these mills are designed to comminute only one or two samples per cycle. In addition, considerable amount of cleaning time between samples is needed to prevent cross-contamination in these small capacity mills.

3.2.8   Finishing

Established analytical techniques including X-ray fluorescence spectroscopy (XRFS) and laser ablation ICP-MS have been developed to analyse samples in the form of round, flat-surfaced disks or pellets. Sample pellets are prepared by milling a representative sub-sample into a fine powder (i.e. 100% less than 30 microns) prior to pressing the powder into a soft metal pellet former, or as a free pellet without the metal former. Pressing into a former is useful for friable powders which do not form viable free pellets. The former, often an aluminium dish or a lead, steel or rubber *O* ring, deforms as the pressing pressure is applied, reducing the problem of pellet fracture as the pressure is released. Pressing using a former may be analytically advantageous (Buhrke *et al.* 1998), and may avoid the need to introduce an adhesive-binding agent to produce a suitable pellet.

Of critical importance in producing high quality pellets is the dwell time used during pressing (i.e. the time period that the powder is held under the full pressing pressure). Many powders which are difficult to press may be pressed successfully using dwell times of one or two minutes; under normal conditions only three or four seconds is required. Equally important is the rate at which the pressing pressure is reduced after pressing; this should be gradual over a few seconds. Pellet fracture will often occur if the pressure is released too quickly. In addition, powders must be as dry as possible to prevent volatilisation of water into the spectrometer during analysis and to ease cleaning of the pressing tool.

Unfortunately, most dried soil-based samples do not produce robust pellets without the addition of a binding agent. The binder is mixed with the sample by milling or, in the case of liquid binders, by hand kneading in a plastic bag or other flexible container to give maximum contact between the binder and the sample. The mixture is then pressed in a highly polished metal die using up to 0.4 Mpa pressure. Careful control of the sample preparation method allows a high productivity of pellets (up to 300 per man day) and good reproducibility in terms of pellet density and smooth surface texture (Buhrke *et al.* 1998).

A large number of binding agents are available. Synthetic wax powder (Van Zyl 1982), methyl methacrylate powder (Ingham & Miles 1995), polyvinyl alcohol, urea and cellulose/polymer mixtures, paraffin and potassium bromide are examples of binders in regular use, and which are suitable for many sample types. The amount of binder added to the sample is normally the minimum amount required to achieve a robust pellet, additions of up to 25% w/w are used for solid binders, with 5–20 drops per 10 g of sample for liquid binders. When only a small amount of sample is available, the binder may be used in the form of a backing layer behind the sample to provide the pellet with sufficient strength to withstand handling and analysis.

3.2.9   Comminution of vegetation

A detailed treatment of the preparation of plant materials falls outside the scope of this chapter. However, many contaminated samples do contain at least some plant material, which may form an important part of the total sample. Small amounts of grass, leaves and roots which regularly pass through initial sieving do not normally constitute a problem, and are dried and milled successfully with other parts of the sample. If larger amounts of vegetation or individual plants, small branches or root crops need to be prepared, they can be easily removed from the sample by hand or during sieving, and depending on the parameters to be determined they may be oven dried, microwave dried, air dried or freeze dried until friable (drying is not suitable for volatile or labile organic parameters). If it is intended to undertake trace element analysis of plant material only, care must be taken to avoid contamination by soil or dust during collection. The washing of freshly cut plant material to remove contamination is not desirable because soluble constituents may be removed from the sample. Comminution of the dried plant or plant/soil mixture using a knife mill, rotor mill, cyclone mill or blender is normally straightforward and produces a finely divided powder of less than 1 mm maximum particle size. Table 3.2 provides a summary guide to the equipment covered in this chapter.

The importation, handling and disposal of plants, seeds, soil, sediments and other materials that may contain plant diseases are often regulated nationally and care must be taken to avoid contravention of these controls.

Table 3.2 Summary guide for application of comminution equipment covered in this chapter

Application	Common names	Mode of operation
Cutting	Diamond saw	Motor driven saw
	Metal shear	Large hand operated cutter
Milling hard or resistant objects (metals, wood, plastics hide, bone)	Granulator	Blades cutting against a mesh
	Disk mill	Rotating toothed disks
	Knife mill	Heavy rotating steel blades
Crushing hard/brittle objects (rocks, slags, ashes, landfill)	Jaw crusher	Angled jaws
	Roller crusher	Counter-rotating rollers
	Disk mill	Rotating parallel toothed disks
	Fly press	Parallel jaws
Coarse milling resistant materials (soils, sediments crusher product)	Tema mill	Milling elements inside an oscillating vessel
	Shatterbox	As above
	Cross-beater mill	Rotating hammers
	Hammer mill	Blades cutting against a mesh
	Knife mill	Rotating blades/fixed hammers
	Pin mill	Parallel rotating plates with interlocking pins
	Soil mill	Rotating plastic brushes inside a metal cylinder
	Mixermill C2000	Heavy milling puck inside a large closed vessel
Milling soft flexible materials (plastics, muscle fibres)	Freezer mills	Impact grinding using a steel ball or impactor in a cylindrical vessel
Fine milling low resistance materials (clays, vegetation, etc.)	Ball mill	Milling balls inside a rotating vessel
	Mixermill	Single milling ball in an oscillating cylinder
	Micronising mill	Agate cylinders in a sealed vibrating vessel

3.3 Processes and problems

Except when sampling directly from the container for labile parameters, preparation methods generally involve between one and four comminution processes and three or four handling processes (e.g. sub-sampling, sieving), each of which has associated problems with respect to protecting the integrity of the sample. Table 3.3 identifies some common preparation processes and problems encountered in the laboratory.

3.3.1 Contamination from preparation equipment

In most routine sample preparation methods used for contaminated soil, the addition of contaminant materials to the samples during preparation is almost inevitable. Sample containers, drying trays, spatulas and other apparatus which contact the samples directly are obvious potential sources of contamination. Water resistant paper or synthetic inserts placed inside the lids of sample containers may leach contaminants including zinc and boron. Syringes or other apparatus used to sample liquids and gases directly from sample containers may cause addition of unwanted materials. Brushing dried soils from drying trays, or any other process involving unnecessary abrasion of the samples against a fixed surface is not desirable.

All samples comminuted using metallic equipment, particularly crushers and coarse milling apparatus, will receive metallic particles abraded from the surfaces of the jaws, rollers or milling elements during comminution. Contamination levels are difficult to predict due to the random motions and impacts occurring during comminution and due to the fact that no two samples have the same resistance to comminution. However, careful selection of sample preparation methods can result in equipment contamination being of little or no importance with respect to the total sampling/preparation/analytical error. To indicate the potential for equipment-derived contamination, 150 g samples of quartz (99.98% SiO_2) a moderately resistant abrasive mineral were milled on a Tema type mill for seven minutes. The data produced by elemental

Table 3.3 Common problems associated with initial sample preparation processes

Process	Problem
Drying	Contamination, loss of PAHs, Hg, phenols, cyanides; low CEC values
Disaggregation	Loss of fine particulates
Coarse crushing (jaw or roller)	Contamination, cross-contamination
Coarse milling (shatterbox, Tema)	Caking, loss of volatile/labile analytes
Sub-sampling	Particle size or shape fractionation
Fine milling	Loss of fine particulates

Table 3.4 Contamination from comminution equipment

Process (trace element preparation)	Contaminant element (ppm)							
	V	Cr	Mn	Fe	Ni	W	Co	Y
Tema mill (silicon nitride)	7[a]	5[a]	<1	698[a]	<1	<1	<2	30
Tema mill (WC + Co vessel)	nd	5	3	188	<1	576[b]	96	nd
Tema mill (agate vessel)	<1	12[a]	2	152[a]	<1	<1	<1	<1
Tema mill (Ni-steel vessel)	2	748	10	1400	2527	<1	<1	<1
Tema mill (Cr-steel vessel)	2	171	10	1100	2	<1	<1	<1
Mixermill C2000 (Cr-steel)	1	387	20	3200	4	<1	<1	<1
Disk-mill (Cr-steel)	5	786	30	6600	8	<1	<1	<1

[a]Deposited during crushing using a chromium-steel jaw crusher.
[b]Implied contamination based on the known composition of the milling vessel.
nd = Not determined.

(XRFS) analysis of the mill product are given in Table 3.4. The feed size was 10 mm (nominal maximum particle size).

Table 3.4 shows typical equipment contamination levels for 150 g quartz samples undergoing coarse comminution, except Mixermill and Disk-mill values which relate to 1.0 kg samples processed to give 100% >250 micron and 100% <1 mm products respectively (i.e. routine milling methods). An investigation of equipment contamination undertaken at the British Geological Survey produced strong indications that the amount of contamination added to samples can be attributed to the following factors (Allen 1998):

(a) Amount of sample loaded into the milling vessel (small sample = higher contamination).
(b) Resistance or abrasiveness of the sample (resistant sample = higher contamination).
(c) Power of the mill (highly dynamic mill = higher contamination).
(d) Milling time (long milling time = higher contamination).
(e) Condition of the equipment (worn or abraded surfaces = higher contamination).
(f) Density of the milling elements (high density mill vessel = higher contamination).

Both resistance and abrasiveness of most soil samples are lower than those of quartz, but contamination introduced into resistant soil samples may approach the levels listed in Table 3.4. Other investigations of potential contaminants have identified elevated levels of Co, Ti, W and Ta after milling quartz chippings in tungsten carbide (Vander Voet & Riddle 1993). Zirconium oxide milling is likely to introduce ZrO_2, HfO_2, MgO, Ba, Ti, Sr and Yb. Sintered carborundum and hard porcelain may introduce Al_2O_3 and SiO_2. Milling in manganese steel or cast iron may contaminate samples with Mn, Fe,

C and Si. Chromium steel milling may add Cr, Mn, Fe and Ni. Blending soft materials (e.g. clays, vegetation) in food mixers may introduce Ti, Cu, Ni or Zn depending on the composition of the blades. Contamination by lubricating oils and greases may produce anomalous data for Li, Mo and C and PAH or SVOC screens. Although this is by no means an exhaustive list the potential for contamination of samples during preparation is apparent.

Despite their insubstantial nature, plant materials are also subject to contamination by sample containers and milling equipment. Paper sample bags may act as sources of contamination including boron (Standing Committee of Analysts 1986). Normally, vegetation is comminuted in high shear cyclone mills or blenders fitted with stainless steel or titanium-coated blades. As with other more resistant materials the amount of contamination from the mill can be related to the abrasiveness of the sample (Allan *et al.* 1999).

Sample preparation for determination of organic compounds presents somewhat different problems (see Section 3.2.2). Volatilisation of material during drying, milling or sample handling, sorption on to synthetic containers (Barcelona *et al.* 1985) or nylon sieve mesh (Standing Committee of Analysts 1986) and liberation through over-milling are all potential problems. Sources of contamination include fillers found in plastic bags and sample containers, wax coatings on container lids, hair, saliva and skin flakes. Careful planning of preparation to avoid contamination by synthetic equipment (e.g. spatulas, cleaning solvent residues, etc.) and loss by volatilisation is essential.

Most contaminated soils and landfill samples contain abrasive particles, and abrasion from sieve meshes may be a significant source of contamination when preparing material for trace level analysis. Metallic sieve meshes (normally stainless steel or brass) are generally considered to be more accurate than nylon because they are more rigid during sieving, particularly when hand sieving using a soft brush. Analytical priorities normally dictate the type of sieve mesh used.

3.3.2   Cross-contamination

This is a very important source of error, because the variations in composition between individual contaminated samples are often large. Comminution equipment is not generally well designed for cleaning and the potential for carry over of powder or any associated liquids from one sample to another is considerable unless careful attention is paid to cleaning between samples. Crushing samples containing humus-rich soil, aggregates bound in bitumen, etc. may cause serious difficulties in preventing cross-contamination. Washing the equipment with a 5% solution of hydrochloric acid (0.5 M) followed by rinsing with cold water and immediate drying may remove most contamination, but crushing quartz or rock chippings between samples may be the only system for preventing trace level carry over. Many high-throughput laboratories employ

Table 3.5 Cross-contamination by metallic copper following repeated milling of cleaning sand

Order of milling	Cu (%)
Sample 1	25.1
Sample 2	14.8
Sample 3	7.0
Sample 4	3.4
Sample 5	2.0
Sample 6	1.0
Sample 7	0.7
Sample 8	0.4
Sample 9	0.2
Sample 10	0.1

large-scale shatterbox type mills using up to 2000 ml metallic milling vessels served by compressed air cleaners to remove residual powder from vessels at the end of each milling cycle. Following removal from the mill vessel by the air stream the powder is collected by a dust extraction system, and in some cases this type of rapid cleaning is augmented by grinding a charge of sand, limestone or other abrasive material to remove any ultra-fine particles staining the surfaces of the vessel.

The potential for carry over from one sample to other samples processed subsequently is illustrated by Table 3.5. A 1 kg sample of mineral concentrate containing copper (35.9%) was milled for seven minutes in a 2000 ml chromium steel vessel loaded on to a Mixermill C2000. Ten charges of 10 mm pure quartz chippings (500 g) were then milled for three minutes to assess the extent of cross-contamination by Cu. No additional cleaning of the vessel was undertaken and the data in Table 3.5 were produced following XRFS analysis of pressed powder pellet samples prepared from the milled product.

Cross-contamination by carry over of air-borne dusts is generally less important as a route for cross-contamination unless dust control measures are inadequate. However, efficient dust control is an essential part of good laboratory practice.

3.3.3   Sub-sampling bias

Many samples contain malleable particles of metal, plastic and rubber, which are not readily comminuted, or by virtue of their shape, density or hardness are resistant to comminuting forces. The best known of these effects is the *nugget effect*. During milling, dense malleable particles (e.g. gold, platinum and soft metal alloys) are deformed into cigar-shaped particles or smeared on to the surface of milling equipment before being rolled into tubular shapes by the repeated oscillations of the milling rings. These deformed particles then sink rapidly within the powder and can cause very large sub-sampling errors even when accepted sampling procedures (e.g. cone quartering or riffling) are followed. Other workers have noted this effect with respect to lead particles and solders (Smith & James 1981). Low density or platey particles including light minerals, shells, foam rubbers, plastics and fragments of vegetation, etc. may become concentrated at the surface or within the surface layers of the sample. Folding the milled powder with a spatula, blending on a tubular ribbon mixer, or roller blending using a container with internal fins to prevent stratification of the powder are useful methods for limiting separation effects before sub-sampling.

The dimensions and shape of the implement used for sub-sampling coarse samples (e.g. dried soils) have been found to introduce sub-sampling bias. Scoops with sides are preferable to flat spatulas or pallette knives because larger particles tend to roll off the heap which is formed when the sample is taken (Smith & James 1981).

Established theoretical treatments have been produced by various workers including Gy (1979), Ingamells (1974) and Visemann *et al.* (1971) describing the factors affecting error in sampling methods. The equation developed by Gy can be used to provide information on the weight of sub-sample needed to provide an analysis within specified error limits, a determination of the analytical error which would be incurred if a given sub-sample weight were taken, and the maximum particle size to which a material should be reduced to provide a suitably representative sub-sample. Gy's equation can be written:

$$M = \frac{Cd^3}{S^2} \tag{3.1}$$

where:

M = mass to be sampled (g),
d = maximum particle size (cm),
C = sampling constant,
S^2 = analytical variance.

By simplifying the theory developed by Gy we can derive a safety rule:

$$W \geq 250\, d^3 \tag{3.2}$$

where:

W = minimum sub-sample size (g),
d = maximum particle size (cm).

This safety rule can be used to construct a simple table broadly indicating the maximum particle size and minimum sub-sample weight limits for *safe* sub-sampling.

Table 3.6 Minimum aliquot size versus maximum particle size

Maximum particle size (cm)	Minimum sample weight (g)
10	250000
5	31250
2	2000
0.2	2

The Gy equation assumes that random samples are taken from a mixture in which all particles have an equal opportunity of being selected (i.e. the sample must be completely homogenous). In addition, it assumes that the statistical errors associated with random sampling follow a Normal distribution (i.e. it is important to eliminate sub-sampling bias caused by poor equipment or an unsuitable sub-sampling method).

Ingamell's equations show how the size of the sample needed to minimise sampling error is dependent on the maximum particle size of the sample, using a sampling constant Ks, the amount of sample required to ensure that the sampling uncertainty will not exceed a desired level assuming a precise analytical method (Vander Voet & Riddle 1993).

The theory put forward by Visemann *et al.* (1971) allows a calculation of sampling error for sub-samples of any weight taken from a segregated bulk sample by relating segregational variance to random sampling variance. The optimal weight of sample may then be defined when random sampling variance is equal in magnitude to segregational variance.

$$S^2_{TOTAL} = \frac{A}{W} + \frac{B}{N} \tag{3.3}$$

where:

S^2_{TOTAL} = variance of the total sampling error,
A = a random variability constant,
B = a segregation variability constant,
W = sample weight,
N = number of increments constituting the sub-sample.

Detailed discussions on the applications and other limitations of Gy's theory are provided by Ottley (1966), and on Ingamell's and Visemann's theories by Smith and James (1981).

All of these theories imply that the best practice is achieved when sampling is done from a well-mixed material. In essence, processing the largest sample possible and ensuring a high comminution ratio at the coarse crushing stage of preparation will significantly reduce the importance of cumulative errors associated with sub-sampling and comminution during sample preparation. For practical purposes, it is perhaps advisable to ensure that any sub-sample taken is significantly larger in size and has smaller maximum particle size than that required by theory.

3.3.4 Sub-sampling bias caused by sieving operations

Many laboratories screen soils at 2.0 mm, removing all oversize material retained on the screen. In most cases this ensures that the majority of samples comprise the original soil, sub-soil or sediment substrate. This <2.0 mm material is then prepared for analysis. Other laboratories routinely screen at 10.0 mm and include many particles which are highly resistant to preparation and which may result in these samples having a significantly different overall composition compared to samples screened at 2.0 mm. Indeed the selection of one or the other of these methods may significantly alter the interpretation of the analytical data (see Chapter 1, Section 5). Normally results are derived from the fraction of the soil remaining after stone removal. However, some customers require analysis and reporting to include the stone fraction. It is important that the client and the analysing laboratory are fully aware of the sample preparation method used when any analytical results are reported.

It is suggested that in the absence of other project aims, which may pre-select the appropriate screening method, one practical solution to this problem is a careful particle-size analysis of bulk samples taken throughout the location of interest. This may then identify whether the >2.0 mm material is relatively localised and therefore unrepresentative of surface contamination, or present throughout the site as a major component which should be included in the samples analysed. Sub-sampling and comminution can then be planned to maintain the integrity of the samples collected.

3.3.5 Loss of fine particulates, volatile and labile components

Some samples containing predominantly hard-brittle or soft-friable components are quickly reduced to fine air-borne dust during initial crushing (e.g. industrial filter residues) and in some cases comminution of the wet sample may be appropriate to limit the loss of fines. To avoid excessive heating during milling, samples containing semi-volatile or labile compounds (e.g. soils, sediments, slags containing hard resistant particles) are normally crushed as finely as possible (to <1 mm) before milling or, alternatively, large particles are removed for separate crushing. This reduces the milling time needed to comminute the sample because the maximum particle size is smaller, preventing heating of the fine more reactive fraction. Metallic milling systems exhibit an increased tendency for heat generation when compared to non-metallic systems, indeed it is necessary for the operator to wear heat resistant gloves when milling resistant samples. Minimising milling time coupled with cooling the vessel after milling each sample is essential if important volatile components are held within a resistant matrix (e.g. slag, concrete or plastic).

Comminution using metallic parallel plate, hammer and cross-beater mills may generate a great deal of frictional heating within the first few seconds of operation. This problem can be overcome by increasing the aperture size in the mill screen, reducing the feed rate or reducing the feed size, although heat generation in this type of milling can be difficult to overcome.

Failure of the crusher or mill to produce a suitable product should be avoided by experimental milling of surplus material (if available). Caking is often the cause of inefficient milling. The sample merely accumulates as an amorphous coating on the surfaces of the equipment, preventing comminution. Further drying normally remedies this condition but some difficult clay-rich soils or samples containing small amounts of viscous liquids (e.g. oil, tar, etc.) may continue to *sweat* during milling even after further drying and disaggregation. In this case the vessel may be cleaned and the samples re-loaded repeatedly until sufficient reduction has occurred. Alternatively, super-cooling in liquid nitrogen or freezing prior to milling may overcome the problem. Caking may also occur as a consequence of over-milling, as happens when size reduction ceases but the mill is allowed to continue running. The fine particles may then begin to agglomerate forming hard flattened aggregates, which eventually combine into a compacted layer within the milling vessel. Experienced operators will often be aware when milling is complete by listening to changes in the sounds produced by the milling vessel.

Samples containing large amounts of hard, resistant material may severely reduce the efficiency of milling, particularly agate milling (and occasionally tungsten carbide and zirconium oxide milling). Particles may remain un-milled regardless of the milling speed or milling time used. Reducing the sample size, by milling the sample as two sub-samples and re-combining them or re-crushing to reduce the maximum particle size normally overcomes this difficulty. Particularly, resistant samples can be comminuted by *pulse milling*, where the sample is re-loaded into the mill ensuring that no material resides beneath the ring or puck. The mill is then operated for approximately fifteen seconds and allowed to come to rest for a few seconds, before milling again for fifteen seconds. It appears that the movements of the mill table are more random and destructive in the early part of the operation before they settle down to a more regular (circular or oval) pattern typical of mixer mills, shatterboxes, Tema and disk mills.

3.4   Quality control

Many laboratories operate quality control (QC) procedures to underpin sample preparation methods. Laboratories processing large numbers of relatively uniform samples, as in a production line environment, or research laboratories processing a smaller number of samples with varied matrices, have effectively the same QC aims. These are to identify contamination and/or cross-contamination events at the earliest opportunity and provide a verifiable record of the passage of individual samples through the laboratory. Typically, samples are segregated on receipt and processed in facilities dedicated to a particular class of material (e.g. ore-grade, high-trace, trace and ultra-trace preparation) or to the analytical technique for which the samples are to be prepared. The preparation of blank samples is also a common practice in which one or more samples of sand, quartz or limestone are prepared with each sample group, and the data generated are often available to customers. Analysis for expected equipment contamination (e.g. Fe, Cr, Ni, Mn, W, Co) has the added benefit of identifying wear in crusher jaws or milling vessels. This is denoted by an increase in contamination resulting from increased residence time (crushers) or roughened surfaces (milling vessels). Dust monitoring provides an indication of the potential for air-borne cross-over from sample to sample and may be essential as a proof of safe working conditions. Monitoring individual staff and general laboratory conditions using light scattering probes and pump-driven collecting apparatus provides useful information on working systems by identifying dusty processes and dusty technicians.

Many projects involve collecting hundreds of samples which are transferred to the preparation laboratory for processing. From these original samples three or four hundred sub-samples may be generated as the material is processed. As a result, sample and sub-sample labelling is of paramount importance. Strict controls on labelling involving designating each sample with a unique identifier and recording the processing of individual sub-samples are highly desirable in all situations, and essential in high-productivity environments. It is important that good communication exists between the customer, preparation technicians and analysts to ensure that labelling errors do not result in inappropriate sub-samples being analysed.

3.5 Good laboratory practice in initial sample preparation

At present there is no universally accepted 'Standard Method for Preparation of Samples from Contaminated Sites' and few laboratories operate strictly comparable sample preparation methods. So it is important that the methods used for initial sample preparation are verifiably *fit for their analytical purpose* and applied by trained staff under an appropriate level of quality control. The design of sample preparation methods is generally constrained by the need to produce powders or finished samples with uniform physical characteristics by the most rapid method, whilst preventing the loss or alteration of any part of the sample by contamination, cross-contamination or chemical reaction.

A clear definition of the analytical requirements is the first consideration. In most cases, the preparation method will be dictated by the tests and analyses needed. The preparation methods needed to service these requirements should then be considered, taking into account the physical nature of *all* the samples involved. Ideally the same methods should be applied to all the samples undergoing preparation for each analytical technique but the nature of the samples may limit the scope for achieving all the required analytical aims. If sample preparation for one analysis technique (e.g. trace level analysis for heavy metals) compromises preparation for other techniques (e.g. determination of Hg and VOCs) this should be identified at the planning stage. Field duplicate samples or laboratory replicate samples may then be taken and, if appropriate, prepared using facilities dedicated to the particular sample type.

Regardless of their purpose, well designed sample preparation methods have common distinguishing features.

(a)   The minimum number of handling, comminution and sub-sampling processes.
(b)   A quality control procedure incorporating sample traceability records and a mechanism for detection of contamination events.
(c)   Written procedures including *what if* guidance, equipment cleaning methods and details of reasons for deviation from routine procedures.
(d)   The cleanest practicable working conditions irrespective of the grade of preparation involved.
(e)   Trained staff.
(f)   Health and safety controls.

3.5.1   Health and safety

Staff employed in sample preparation laboratories are arguably at a greater risk of exposure to hazardous samples than either the sample collectors or analysts. This is primarily because preparation technicians are exposed to larger volumes of dusts and/or aerosols for longer periods of time compared to other project workers. The protection of such staff is important and should be considered at the project planning stage i.e. before samples are collected. A detailed discussion of this topic would be unlikely to account for variations in national regulations; however, potentially dangerous samples (i.e. radioactive or pathogenic materials) should be clearly notified by the collector and appropriate precautions communicated to all laboratory staff including those who may be indirectly exposed. Screening for hazardous samples is normally undertaken as part of sample registration but the importance of early warning information from customer/collector cannot be over-stressed. Appropriate training of staff and continuous use of protective equipment graded to cover major spillages or other uncontrolled hazards are advisable. Potential inhalation of asbestos fibres from *unknown* samples is a very serious risk that must be addressed.

3.5.2 Provisional protocols

Using the previous text as a guide it is possible to propose provisional protocols for preparing contaminated samples subject to the analytical focus of the site investigation.

Suggested protocol for non-volatile/non-labile components

Receipt and temporary storage	
Inspection and sample registration	Confirmation of preparation processes to be used with agreement of analyst and customer.
	Entry of sample identifiers onto database or LIMS.
	Extraction of gases, pore waters, leachates or *as received* sub-samples as required.
Homogenisation	By hand or mechanical.
Sub-sampling	For reference sample (if required by customer or analyst).
Drying	To *constant* weight at <35–105°C (actual drying temperature dependent on analytical requirements).
Disaggregation	Mortar and pestle, or mechanical (if needed).
Sieving	Metallic or nylon test sieve mesh.
	If necessary, remove >2 mm or >10 mm material as appropriate.
	Comminute oversize if appropriate.
Comminution	Jaw or roller crusher (if required) down to 3 mm (nominal).
Sub-sampling	(if necessary) Cone-quarter or riffle (largest sub-sample practicable).
Coarse milling	Ring mill, Shatterbox or Tema mill (down to 200 microns, nominal).
Sub-sampling	Cone-quarter or riffle for analytical dissolution techniques suitable, for e.g. ICP-AES, AFS-HG (excess material to archive or customer).
Fine comminution	Ball mill or other mill (down to 40 microns) with or without binder.
Sub-sample	(if necessary) Cone-quarter, riffle or dipping technique (excess powder to archive).
Finishing	Pellet preparation, ultra-fine milling.

Suggested protocol for semi-volatile components

Receipt, inspection and registration as above.

Homogenisation	By hand or mechanical (if required).
Sub-sampling	Reference sample if required by customer/ analyst, or for immediate analyses (PCB's, PAH's, etc.).
Drying	To *constant* weight by air drying at <30°C (or at ambient temperature, if higher).
Disaggregation	(if needed).
Sieving	Metallic or nylon test sieve. Remove >2 mm or >10 mm as appropriate. Comminute and re-combine oversize if required.
Comminution	Jaw or roller crusher (as fine as possible but <2 mm in any case).
Sub-sampling	(if necessary) Cone-quarter or riffle (largest sub-sample practicable).
Coarse milling	Shatterbox or Tema milling (minimum milling period – freezer mill or cooled milling vessel).
Sub-sampling	Cone-quarter, riffle for analytical dis- solution techniques (excess material to archive or customer).
Fine comminution	Ball mill (minimum milling period).
Sub-sample	(if necessary) Cone-quarter, riffle or dipping technique (excess powder to archive).
Finishing	Pellet preparation (if required).

Volatile/labile components

See Standing Committee of Analysts (2003) and ISO/FDIS 14507 (2002).

References

Allan, A.M., Blakeney, A.B., Batten, G.D. & Dunn, T.S. (1999) Impact of grinder configuration on grinding rate, particle size, and trace element contamination of plant samples. *CA Selects: Trace Element Analysis*, Issue 23, 2.

Allen, M.A. (1998) Comminution of geological samples using syalon based milling vessels: comparison with established methods. *Analytical Communications*, **35**, 75–78.

Barcelona, M.J., Helfrich, J.A. & Garske, E.E. (1985) Sample tubing effects on groundwater samples. *Analytical Chemistry*, **57**, 460–464.

Broton, D. (1996) *Practical Aspects of Sampling and Preparing Powders for XRF Analysis* (Unpublished conference notes).

Buhrke, V.E., Jenkins, R. & Smith, D.K. (1998) *Preparation of Specimens for X-ray Fluorescence and X-ray Diffraction Analysis*, Wiley-VCH, p. 188. ISBN 0471194581.

Creedy, D.P. (1986) Methods for the evaluation of seam gas content from measures on coal samples. *Mining Science and Technology*, **3**, 141–160.

Gy, P. (1979) *Sampling of Particulate Materials – Theory and Practice*, Elsevier, Amsterdam. ISBN 0444418261.

Ingamells, C.O. (1974) *Talanta*, **21**, 141.

Ingham, M.I. & Miles, P.H. (1995) *An Investigation of the Suitability of Elvacite, used as a Liquid Binder, in Pressed Powder Samples for Analysis by X-ray Fluorescence Spectrometry*, British Geological Survey Technical Report Number WI/94/5.

ISO 11464 (1994) *Soil quality – Pre-treatment of samples for physico-chemical analysis*.

ISO/FDIS 14507 (2002) *Soil quality – Pre-treatment of samples for determination of organic contaminants*.

Mutter, W. (1997) Sample division as a basis for modern analytical methods, *International Laboratory News* (June Edition).

Ottley, D.J. (1966) Pierre Gy's sampling rule. *Canadian Mining Journal*, **87**, 58–65.

Popp, M., Lied, M., Meyer, A.J., Richter, A., Schiller, P. & Schwitte, H. (1996) Sample preservation for determination of organic compounds: microwave versus freeze-drying. *Journal of Experimental Botany*, **47**(303), 1469–1473.

Schmidt, M.W.I., Knicker, H., Hatcher, P.G. & Kogleknabner, I. (1997) Does ultrasonic dispersion and homogenization by ball milling change the chemical structure of organic matter in geochemical samples? A CPMAS C-13 study with lignin. *Organic Chemistry*, **26**(7–8), 491–496.

Smith, R. & James, G.V. (1981) *The Sampling of Bulk Materials, Royal Society of Chemistry Analytical Sciences Monographs*. ISBN 085186810X.

Standing Committee of Analysts (1986) *The Sampling and Initial Preparation of Sewage and Waterworks Sludges, Soils, Sediments, Plant Materials and Contaminated Wildlife prior to Analysis*, second edition, HMSO. ISBN 0117518859.

Standing Committee of Analysts (2003) *Pre-treatment of Contaminated Land Samples for Chemical Analysis*, Standing Committee of Analysts, Environment Agency England and Wales.

Vander Voet, A.H.M. & Riddle, C. (1993) *The Analysis of Geological Materials Volume 1: A Practical Guide*, Miscellaneous Paper 149, Ontario Geological Survey, Canada.

Van Zyl (1982) Rapid preparation of robust pressed powder briquettes Containing a styrene and wax mixture as binder, *X-Ray Spectrometry*, **11**(1), Heyden and Son Ltd.

Visemann, J., Duncan, A.J. & Lerner, M. (1971) *Materials Res. and Standards*, **11**, 32.

White, R.E. (1987) *Introduction to the Principles and Practice of Soil Science*, 2nd edn. Blackwell Scientific Publications, Oxford, UK. ISBN 063201606X.

4 Metal analysis

Patrick Thomas

4.1 Introduction

Many metal pollutants (e.g. mercury, cadmium, lead, nickel and zinc) are hazardous to human health and terrestrial ecosystems. The determination of metals in contaminated soils may be carried out for a variety of reasons, such as measurement of total or defined extractable content, which provides knowledge of soil components in respect of changes in soil composition produced by natural or anthropogenic contamination. In addition, the analysis may be designed to investigate and categorise degrees of contamination of land for current or intended use. An environmental survey of a polluted site may be undertaken to obtain information about the nature, quantity, distribution and behaviour of contaminants and, if necessary, to select the most appropriate use of the site (Hester & Harrison 2001; Kibblewhite 2001).

This chapter is designed to provide an analyst's prospective on commonly used techniques for the determination of all relevant metals, mainly after wet acid digestions. Other dissolution methods are described briefly, problems with the analysis of the digestion solution are highlighted and pragmatic choices are proposed. The analysis of the total major components (SiO_2, Al_2O_3, Fe_2O_3, CaO, MgO, TiO_2) and on-site analytical methods used for site investigations are not covered in this chapter. Sulfur analysis is discussed briefly in the case of contaminated agricultural land and a brief view of speciation analysis on soil samples of relevant toxic element species, such as cadmium, nickel, chromium, arsenic, selenium, mercury and organotin compounds, is given.

In this chapter, two aspects of analysis are considered. Firstly, the digestion step of contaminated soil samples is discussed. Secondly, the various techniques of determination of all relevant elements are considered. Important criteria for selecting an analytical technique include detection limits, analytical working range, sample throughput, cost, interferences, ease of use and the availability of proven methodology. These criteria are discussed below for the major atomic spectroscopic techniques.

4.2 General discussion

In practice, the amount of contaminants in a given soil is unknown and can cover a wide range of concentration. Nevertheless, the analyst has to match the method with the analyte concentration. Before processing reliable data, several steps are needed to collect information for soil characterisation; factors affecting this information are explained below.

Sampling of potentially contaminated soil from polluted areas is intended to provide data for assessment of whether the pollution has caused or may cause environmental problems. When carrying out environmental soil investigations or surveys, the analyst is faced with sampling heterogeneous material, involving a substantially greater degree of uncertainty than laboratory analysis of the sample itself. Sampling is probably the step of greatest importance to the reliability of a site investigation. However, it must be underlined that one of the crucial steps in attaining comparable data, except for the sampling, is the sample pre-treatment (drying, sieving, grinding, mixing...). These steps are fully discussed in Chapter 3.

Nowadays, there are no nationally and, may be only very few, internationally recognised standard methods for analysis of toxic metals in polluted soil samples. An exception is the USEPA publication *Test Methods for Evaluating Solid Waste, Physical/Chemical Methods* (SW 846 EPA). There are many methods described for sediment analysis, but in European countries, the Nordic guidelines for chemical analysis of contaminated soil (Kaerstensen 1996) is the only technical report dealing with the analysis of potentially polluted soil. However, recently, the BRGM's Scientific and Technical Centre published a general guide on analysis of polluted soils (Jeannot *et al.* 2001). Usually, analytical methods normally applied to solid samples from contaminated land sites are based on recommended methods for water, wastewater and sludges, or agricultural soils. However, chemical analysis of polluted soil samples can be difficult because of interferences due to the complex soil matrix of this kind of sample (e.g. mixture of elements/pollutants at high concentrations (e.g. Al, Ca, Fe) and mixtures of organic compounds such as PAHs, PCBs, hydrocarbons etc.). As previously stated, metals may be determined satisfactorily by a variety of methods, with the choice often depending on the aim of the study, the precision and sensitivity required.

In publications and reports, analytical methods used are not always specified, and results are often presented as *total analysis* data. In many cases, a so-called *partial analysis* method has been used, where the sample has been analysed after digestion with a strong acid solution in conjunction with a hot plate, boiling device or microwave heating system. The decision to use a total (e.g. $HF/HNO_3/HClO_4$) or partial digestion technique (*aqua regia*) is still a source of discussion. Total digestion will dissolve heavy metals, bound to silicates, completely (extraction efficiency close to 100%) but partial digestion will dissolve heavy metals, not firmly bound to silicate phases (extraction efficiency will depend on the element and matrix). This is a problem when data are used in mass balance calculations, and if results based on partial digestion of silicate matrix are used for such purpose, there is a risk of committing errors by underestimation of the amount of element. It is therefore important to state clearly which kind of digestion was used. When partial digestion techniques are used, the results must not referred to as *total analysis*.

However, the advantages of analysis after a partial digestion with a strong acid or a mixture of strong acids are that the methods have been widely used, they generally are reproducible, they are comparatively simple to perform and that they can be performed on a relatively large solid sample. The aqua regia digestion method is empirical and it might not release elements completely. Nevertheless, the data for element concentrations obtained by aqua regia extraction can be useful for an approximate knowledge of the total element content in soil. This is especially true when the soil sample is affected by anthropogenic contamination where potentially toxic elements have been evaluated. In such soils, an easier extractability of the contaminant compared to the element bound into the silicate matrix can be expected (Hjelmar 1999). It is generally accepted for risk assessment purposes that commonly used aqua regia digestion or nitric acid digestion techniques are fit-for-purpose for assessment of contaminated land. Any elements not detected by this procedure are very unlikely to constitute a significant health risk.

There are three main atomic spectroscopic techniques that are used for the analysis of acid digests: atomic absorption spectrometry (AAS), inductively coupled plasma-atomic emission spectrometry (ICP-AES) and atomic fluorescence spectrometry (AFS). Of these, AAS and ICP-AES are the most widely used. Our discussion will deal with these techniques and also an affiliated technique, inductively coupled plasma mass spectrometry (ICP-MS).

A number of techniques are available for the analysis of metals in soil samples; the choice will depend on the desired concentration range and the type of metal. Generally, in soil and waste samples, the following techniques are recommended and used (Hoenig 1998):

1. The most commonly utilised is ICP-AES for the determination of high and trace levels of at least 30 elements. Common examples are the following elements: Ag, Al, As, B, Ba, Be, Ca, Cd, Co, Cr, Cu, Fe, K, Li, Mg, Mn, Mo, Na, Ni, P, Pb, S, Sb, Se, Si, Sn, Sr, Ti, Tl, V and Zn.
2. Alternatively, flame atomic absorption spectrometry (FAAS) is used for the determination of high and trace levels of Ag, Ca, Cd, Co, Cr, Cu, Fe, Mg, Mn, Ni, Pb, V and Zn.
3. Electrothermal atomic absorption spectrometry (ETAAS) also called graphite furnace atomic absorption spectrometry (GFAAS) for the

determination of trace and ultra-trace levels of Ag, As, Be, Cd, Co, Cr, Mo, Ni, Pb, Se, Sb and Tl.

4. Hydride generation atomic absorption spectrometry (HG-AAS) or atomic fluorescence spectrometry (HG-AFS) for the determination of trace levels of As, Bi, Ge, Sb, Se, Sn and Te.

5. Cold vapour atomic absorption spectrometry (CV-AAS) or atomic fluorescence spectrometry (CV-AFS) for the determination of Hg.

6. Inductively coupled plasma mass spectrometry (ICP-MS) now provides better detection limits for important elements (e.g. Cd, Pb and Tl) and larger simultaneous determinand suites than ICP-AES.

4.2.1 Overview of digestion methods

Before going further, it seems very important to keep in mind the problem of sampling polluted soils in representative manner. The aliquot actually analysed (typically 0.5–3 g) can relate to many tonnes or cubic meters of the actual site being investigated (see also Chapter 1). Usually a sample is dried, reduced and sieved so that the test sample portion is representative of the field fraction. Unrepresentative sampling can have a marked effect on analytical results; consequently these preliminary steps must be under tight control.

In practice, all modern analytical instrumental techniques of atomic spectroscopy and all common devices for sample introduction associated with these instruments are designed entirely for analysing solutions. That is why they require the conversion of the soil sample into a liquid form, the digestion step. The digestion of samples is a critical step in the analytical process (often the most laborious and time-consuming, see Fig. 4.1) and yet is often overlooked.

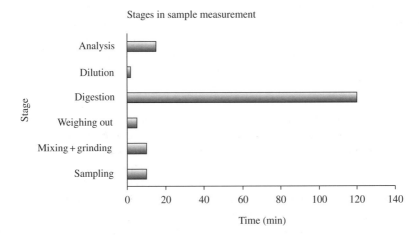

Figure 4.1 Digestion takes longer time than other analysis stages.

The total sample preparation time is the limiting factor for productivity of the analysis.

In addition, errors in the analysis can often be traced to the digestion step, at which conventional methods may, sometimes, lead to non-reproducible results. Typically, errors of up to 30% can be attributed to this sample processing stage (Majors 1991). The digestion process involves the destruction of sample matrices to leave a clear (particulate-free) homogeneous solution for subsequent analysis by atomic spectroscopic techniques. Inorganic samples, such as soils, are digested simply by the dissolution of inorganic materials with acids, such as *aqua regia*. Nearly all methods of final determination of metals, especially routine methods, first require digestion of the soil sample.

Generally, two major groups of decomposition procedures may be identified dry ashing procedures and wet digestion procedures (Hoenig and Thomas 2002).

4.2.2   Dry ashing

The method is simple, and large sample series may be treated simultaneously. Dry oxidation or dry ashing may be used to remove organic matter from soil samples. The soil sample is weighed into a suitable crucible, heated at relatively elevated temperatures (450–550°C) in a muffle furnace with, sometimes, ashing aids to speed up or improve the procedure. Commonly used ashing aids are magnesium nitrate, sulfuric acid and nitric acid. The residue is dissolved in a minimum quantity of appropriate acid (nitric acid, aqua regia or perchloric acid and hydrochloric acid).

However, the dry ashing method cannot be applied if volatile elements such as Hg, As, Sb, Se and Cd are to be determined since they may volatilise during the ashing process. Due to potential volatilisation or losses of analyte into ceramic or silica ashing vessels, the dry ashing procedure is not recommended. However, a dry ashing step has been recommended before carrying out digestion with hydrofluoric and perchloric acid if large amounts of organic matter are present in the soil. This is according to a recent ISO standard document connected with the determination of total trace element content (ISO 14869-1 (2001)).

4.2.3   Wet digestion

Wet digestion is the most common method of dissolving soil samples, and involves the use of a mineral acid or a mixture of mineral acids. Acid digestions are usually carried out either in glass or PTFE vessels. Many different methods of wet digestion are thus available for the determination of the content of trace elements in soil. In the following discussion, the principles, advantages and drawbacks are described for different types of digestion methods.

The selection of acid is a critical decision, given that the obtained solution must be compatible with the analytical method and metals being analysed (Hoenig & De Kersabiec 1996).

Table 4.1 Acids and reagents used in conjunction with nitric acid for soil sample preparation

Acid	May be helpful for	Not recommended for
HCl	Sb, Ru, Sn, Mo, Ba	Ag, Th, Pb
HClO$_4$	Organic materials	
H$_2$O$_2$	Organic materials	
HF	Siliceous materials	

Nitric acid will digest many common metals, but not aluminium and chromium; it is also useful for organic substances. Nitrate is an acceptable matrix for both flame and graphite furnace atomic absorption and the preferred matrix for ICP-MS (Jarvis *et al.* 1992). Therefore, some soil samples may require addition of perchloric, hydrochloric or hydrofluoric acid for complete digestion. These acids may interfere in the analysis of some metals, and all provide a difficult matrix for both graphite furnace and ICP-MS analysis. Table 4.1 provides a guide in determining which acid could be used in addition to nitric acid for complete digestion.

Hydrofluoric acid is mainly used to decompose silicate in soil samples, where silicon is not to be determined and a true total analysis is required. The complete removal of hydrofluoric acid is essential for the success of the subsequent analysis because the fluoride complexes of several cations are very stable. Hydrofluoric acid is often used in conjunction with other acids, for example, in ISO 14869-1, a digestion with a mixture of hydrofluoric acid and perchloric acid is used for complete dissolution of many elements in soil samples. Silicates are decomposed and form volatile SiF$_4$. During the evaporation step, perchloric acid forms soluble perchlorate salts. After evaporation to near dryness, the residue is dissolved in dilute hydrochloric acid. The removal of the last traces of SiF$_4$ from the soil sample is sometimes difficult and quite time consuming. If the HF is not removed, severe corrosion of the nebuliser, spray chamber and torch will occur unless expensive HF resistant materials are used. Perchloric acid is mainly used not as an acid but as an oxidising agent to break up organic material. For soil samples, this acid is used in conjunction with hydrofluoric acid for the extraction of total trace elements in the ISO standard 14869-1. Because of its potentially explosive character, extreme care must be taken. Explosions may occur when hot concentrated acid comes in contact with a significant amount of organic matter or easily oxidised inorganic substances. This is the reason why special precautions are taken, such as the determination of the total organic content of the soil sample before performing this digestion procedure.

Hydrochloric acid dissolves many inorganic salts (carbonates and phosphate) and some oxides. Usually, hydrochloric acid is never used alone in soil analysis. Hydrochloric acid is mainly used together with nitric acid to form aqua regia mixture, a mixture containing HCl:HNO$_3$ (v/v = 3:1). Aqua regia dissolves most metals including noble metals (Ag, Pt, Pd). This acid mixture is well known and is commonly used for partial digestion, for example, the extraction of trace elements soluble in aqua regia (ISO 11466). For normal contaminated land site investigations, this digestion method will effectively give *total metals* suitable for risk assessments and any remaining metal will be firmly bound with the silica matrix and totally unavailable under normal environmental conditions.

4.2.4   Heating devices

Basically, three ways are used for carrying out the wet digestion procedure. The first one is a very conventional device such as conical flask and refluxing device. This is mainly used for aqua regia digestion of a large amount of soil (3 g dry matter) as detailed in the ISO standard 11466 this apparatus can be applied to one sample or a batch of samples.

Until recently, most laboratories were still using hot plate as their only method of sample preparation, a technique that, except for its sources of energy, dates back to the age of alchemy. Because of the indirect nature of heating in hot plate digestions, attainment of the final chemical conditions is relatively slow (Kingston & Walter 1992; Kingston & Haswell 1997).

Hot plate or sand bath digestions are limited by the boiling point of the acid. Recently, some improvements have been made in order to increase the efficiency of the heating process in term of temperature regulation and also in order to increase the number of samples to be digested by the use of a special hot block capable of digesting up to 50 samples in suitable tubes (typically ~50 ml). Electric powered graphite heating blocks are also available with a very good uniform heating across the heated block area. Nowadays, microwave digestion offers an alternative that promises significant saving in time and better reproducibility of the digestion procedure, avoiding losses of volatile components (e.g. Hg, As, Se...).

Microwave digestion considerably speeds up the heating process, often reducing sample digestion time 10-fold even if the vessels must be left to cool down for 15–20 minutes or less if water-cooled before opening. There are two approaches to microwave digestion: the open system (non-pressurised) and the closed system (pressurised). These two methods of heating were included in a recent European Standard (2000) connected with the characterisation of sludges – extraction methods of trace elements soluble in aqua regia (European Standard EN 13346). They have different characteristics and applications, which is why there are so many different models available for each category.

Open systems use focussed microwaves, which are guided on to the sample. Since pressure is not required to achieve high temperatures, non-sealed vessels can be used, with the benefit of being able to digest large amount of sample using a large volume of acid as indicated in the ISO standard (ISO 11466).

A significant advantage, in terms of flexibility and safety, is that acids can be added to the vessels at any stage of the heating process. Besides, it can be used as an alternative term of energy source in order to reduce the sample size, the reagent requirements, the risk of losses, and digestion times (Thomas 1996). However, the open microwave system is not suitable for carrying out total digestion of soil samples including HF and $HClO_4$ with PTFE vessels; several problems may occur during the final evaporation step leading to damage of the PTFE vessels.

The closed system is a microwave oven containing a sealed vessel. In a microwave closed-vessel system, the soil sample and the acid are added to a microwave transparent vessel made of fluorocarbon polymer (PFA or TFM). The vessel is closed and safety contained. When microwave power is applied the pressure in the vessel increases, with a corresponding increase of the boiling point of acid, thus greatly reducing digestion times, to 15–20 minutes from several hours, for almost all kinds of soil samples. Therefore, the reagent combination (nitric acid and hydrochloric acid) results in greater pressures than those resulting from use of only nitric acid. These higher pressures necessitate the use of vessels having higher pressure capabilities. This is the most popular technology for the digestion of inorganic samples, such as those in environmental analysis (e.g. soils, wastes and sludges), because of its high sample throughput and reliability (see USEPA microwave digestion methods 3051A, or 3052). Generally, when carrying out total digestion including hydrofluoric acid, it is necessary to add boric acid at the end of the digestion procedure. In this case, the addition of large amount of boric acid increases the percentage of dissolved salts in the final solution. So, to compensate for matrix effects between samples and standards, a matching solution is required and should contain hydrochloric acid and boric acid. In this case, this is a limitation for a further determination by ICP-MS because highly dissolved solids may cause interface cone blockage and the formation of polyatomic ions from solvent matrix and plasma gases (Gray 1993).

To summarise, microwave sample preparation reduces blank contribution from environmental exposure, reagent use and losses from evaporation, and in addition, reduces sample preparation times. It also offers a more reproducible method for duplicating metals determination, both within a laboratory and between laboratories (Lamble & Hill 1998). But, in the case of contaminated soil sample digestions, due to surface effects of PTFE, a microwave-cleaning step is highly recommended between each bath of samples in order to avoid any memory effects (carry over contamination) from previous digested samples.

Nearly all, especially routine, methods for determination of metals, require digestion of the soil sample. The choice of the digestion technique should take into account the objective of the final determination; incomplete digestion procedures, which require less time and labour, are often acceptable for site investigation. Even though partial procedure does not release elements tightly bound in minerals, assuming that the tightly bound elements are of minimal environmental concern as the metals remaining insoluble after the digestion are not likely to become available, the use of aqua regia or nitric acid is considered justifiable. In Table 4.2, references are given to different relevant standard methods based on destruction of soils and waste matrices.

4.3   Overview of instrumental methods of analysis

4.3.1   What is atomic absorption?

Atomic absorption is the process that occurs when a ground state atom absorbs energy in the form of light of a specific wavelength and is elevated to an excited state. The amount of light energy absorbed at this wavelength will increase if the number of atoms of the selected element in the light path increases. The relationship between the amount of light absorbed and the concentration of analyte present in known standards can be used to determine unknown concentrations by measuring the amount of light they absorb. The instrument readout, usually a computer, can be used to display concentrations directly. The basic instrumentation for atomic absorption requires a primary light source, an atom source, a monochromator to isolate the specific wavelength of light to be used, a detector to measure the light accurately, electronic devices to treat the signal and a data display or logging device to show the results. The light source normally used is a hollow cathode lamp, but in some cases the use of an electrodeless discharge lamp or a boosted discharge hollow cathode lamp for elements such as As, Se, Sn, Sb, Te and Tl is recommended when the resonance spectra of the element is in the far ultra violet region where instrumental efficiency is reduced, and also for determination at or near the detection limit. In some cases, a 10-fold improvement in detection can be achieved.

The atom source used must produce free analyte atoms from the sample. The source of energy for free atom production is heat, most commonly in the form of an air-acetylene or nitrous oxide-acetylene flame. The sample is introduced as an aerosol into the flame. The flame burner head is aligned so that the light beam passes through the flame, where the light is absorbed. Flame atomic absorption (FAA) was commercially introduced in 1961 (Welz & Sperling 1999). It provides specificity, reduced sample manipulation, and application to a wide variety of elements. Since then, atomic absorption has been expanded to include graphite furnace atomisation and also cold vapour and hydride generation techniques. Other related techniques such as inductively coupled plasma atomic emission spectrometry (ICP-AES) and inductively coupled plasma mass spectrometry (ICP-MS) have been developed based on similar spectroscopy principles as shown in Figure 4.2 and will be discussed in this chapter.

Table 4.2 Overview of standard and approved draft standard methods for soil, waste and sludge samples

Method	Soil sample size	Approved or recommended elements	References
Total digestion by acid mixtures			
ISO standard: Soil Quality – Determination of trace element content. Part 1: Digestion with hydrofluoric and perchloric acids	0.25 g sample	Al, Ba, Cd, Cs, Cr, Co, Cu, Fe, K, Li, Mg, Mn, Na, Ni, P, Pb, Sr, V, Zn	ISO 14869-1 (2001)
EN standard: Characterisation of waste. Microwave assisted digestion with hydrofluoric, nitric, and hydrochloric acid mixture for subsequent determination of elements in waste	0.2–0.5 g sample	Al, Sb, As, B, Ba, Be, Cd, Cr, Co, Cu, Fe, Pb, Mg, Mn, Hg, Mo, Ni, P, K, Se, Ag, S, Na, Sr, Sn, Te, Ti, Tl, V, Zn	PrEN 13656 Formal vote (2001–06)
USEPA 3052: Microwave assisted acid digestion of siliceous and organically based matrices	Up to 0.5 g sample	Al, Sb, As, B, Be, Cd, Ca, Cr, Co, Cu, Fe, Mg, Mn, Hg, Mo, Ni, K, Se, Ag, Na, Sr, Tl, V, Zn	USEPA FIN.UP.III IIIA(4/1998)
Partial digestion by acid mixtures			
ISO standard: Soil Quality – Extraction of trace elements soluble in aqua regia	3 g sample	Cd, Cr, Co, Cu, Pb, Mn, Ni, Zn	ISO 11466 (1995)
EN standard: Characterisation of waste. Digestion for subsequent determination of aqua regia soluble portion of elements in waste	0.2–0.5 g sample	Al, Sb, As, B, Ba, Be, Cd, Cr, Co, Cu, Fe, Pb, Mg, Mn, Hg, Mo, Ni, P, K, Se, Ag, S, Na, Sr, Sn, Te, Ti, Tl, V, Zn	PrEN 13657 Formal vote (2001–06)
EN standard: Characterisation of sludge	0.5–1 g sample	Cd, Cu, Cr, Ni, Pb, Hg, P, Zn	EN 13346
USEPA 3050 B: Acid digestion of sediments, sludges and soils	1 g sample	Al, Ag, As, Ba, Be, Ca, Cd, Cr, Co, Cu, Fe, K, Mg, Mn, Mo, Na, Ni, Pb, Se, Tl, V, Zn	USEPA FIN.UP.III (12/1996)
Partial digestion by acid nitric			
USEPA 3051A: Microwave assisted acid digestion of sediments, sludges, soils and oils	Up to 0.50 g sample	Al, Ag, As, B, Ba, Be, Ca, Cd, Cr, Co, Cu, Fe, Hg, K, Mg, Mn, Mo, Na, Ni, Pb, Sb, Se, Sr, Tl, V, Zn	USEPA DRAFT UP-IVA (1/1998)
Swedish Standard: Soil analysis – Determination of trace elements in soils – extraction with nitric acid	Up to 5 g sample	Al, Ag, As, B, Ba, Be, Bi, Ca, Cd, Cr, Co, Cu, Fe, K, Li, Mg, Mn, Mo, Na, Ni, P, Pb, S, Sb, Se, Si, Sn, Sr, Ti, V, W, Zn, Zr	Swedish Standards Institution (1997)
Danish Standard: Determination of metals in water, sludge and sediments	Up to 1 g sample	Ca, Li, Mg, Mn, Ni, Fe, Cr, Cu, Cd, Co, Pb, Zn	Danish Standard (1982, 1990)

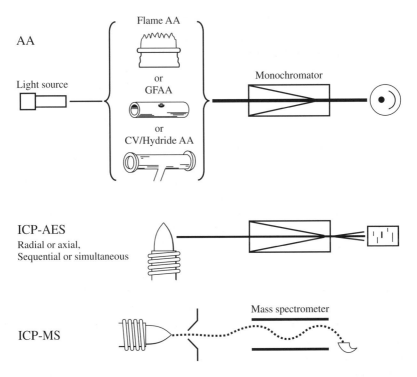

Figure 4.2 Atomic spectroscopic techniques. (Adapted from: Grosser & Schneider 1996.)

4.3.2 Flame atomic absorption spectrometry

Flame atomic absorption spectrometry (FAAS) proved to be a simple, low cost, reliable way of providing metal analysis for hundreds of different applications. In its implementation, the choices the analyst need make are few and include the element which is to be determined, what spectral bandpass to use, what type of flame to use (air-acetylene or nitrous-oxide-acetylene), and whether the flame should be lean or rich. Today, most of these parameters are automated. With modern instruments, all that the analyst has to do is insert the lamp, ignite the flame and recall a method. High solids burner heads and high sensitivity nebulisers are available and extend the capability of the systems for more challenging complex soil extract matrices. Probably the biggest benefit of FAAS is the wealth of experience that exists in using the technique. The relatively few interferences have been thoroughly investigated and solutions verified for the more common ones, these interferences are fully described by Cresser (1994).

It is easy to find a method for most types of matrices and the element of interest in standard methods publication, such as those from the American

Society for Testing and Material (ASTM), US Environmental Protection Agency (USEPA), International Organisation for Standardisation (ISO), European Committee for Standardisation (CEN), Standard Methods for Water and Wastewater and Deutsches Institut für Normung (DIN).

But there is no specific FAAS standard method specially evaluated or approved for polluted soil samples; there is the EPA methods (SW-846 EPA (2000)) or one ISO standard (ISO 11047). This standard is connected with the determination of several metals in aqua regia extracts. It provides a method for the determination of Cd, Cu, Co, Pb, Mn, Ni and Zn by flame and electrothermal atomic absorption spectrometric methods.

Flame atomic absorption used with an autosampler and on-line dilution device, provides a good sample throughput when analysing a large number of samples for a limited number of elements (no more than 12 elements in the same analytical sequence when using multi-element lamps). A typical determination of a single element requires less than two minutes. However, flame atomic absorption requires specific light sources and optical parameters for each element to be determined and may require different flames (air-acetylene or nitrous oxide-acetylene) for refractory elements (Al, Ba, Ca, Cr, Mo, Si, Ti, V) and also different flame stoichiometries. In automated multi-element flame atomic absorption systems, all samples are normally analysed for one element, the system is then automatically adjusted for the next element, and so on. As a result, even though it is frequently used for multi-element analysis, flame is considered to be a single element technique. The analytical working range can be viewed as the concentration range over which quantitative results can be obtained without having to re-calibrate the instrument, assuming that the instrument is working in an approved calibration range. For flame, the analytical working range depends on the instrument sensitivity but generally covers about two to three orders of magnitude.

To summarise, FAAS is very easy to use. Interferences are known and can be controlled. Extensive application information is also readily available. Its precision makes it an excellent technique for the determination of a number of commonly analysed elements at higher concentration in polluted soil samples. Its main drawback is its speed in relation to multi-element techniques such as ICP-AES and ICP-MS. Where direct-aspiration flame atomic absorption technique does not provide adequate sensitivity, reference is made to specialised techniques (in addition to graphite furnace procedure) such as the gaseous-hydride method for arsenic, antimony and selenium and the cold-vapour technique for mercury.

Sensitivity is a term used in atomic absorption spectrometry to indicate the concentration that will cause an absorption of 1% (= 0.0044 absorbance units) of the hollow cathode resonance line radiation used for the determination. An example of typical instrumental sensitivity and detection limits for commercially available instruments is given in Table 4.3.

Table 4.3 Atomic absorption spectrometry: indication of detection limits and sensitivity obtainable by flame and by furnace techniques (by permission of Varian Analytical Instruments)

Metal	Wavelength nm	Type of flame	Direct aspiration			Graphite furnace	
			Detection limit (mg/L)	Sensitivity (mg/L)[a]	Sensitivity (mg/kg)[b]	Detection limit (µg/L for 20 µl injected)	Sensitivity (mg/kg)[b]
Aluminium	309.3	N_2O/C_2H_2	0.02	0.4	40	0.06	0.006
Antimony	217.6	Air/C_2H_2	0.03	0.15	15	0.50	0.05
Arsenic	193.7	N_2O/C_2H_2	0.04	0.6	60	0.22	0.02
Barium	553.6	N_2O/C_2H_2	0.008	0.18	18	–	–
Beryllium	234.9	N_2O/C_2H_2	0.0006	0.01	1	0.01	0.001
Bismuth	306.8	Air/C_2H_2	0.03	0.2	20	0.29	0.03
Cadmium	228.8	Air/C_2H_2	0.001	0.006	0.6	0.009	0.0009
Calcium	422.7	N_2O/C_2H_2	0.001	0.004	0.4	–	–
Chromium	357.9	Air/C_2H_2	0.004	0.06	6	0.038	0.004
Cobalt	240.7	Air/C_2H_2	0.004	0.04	4	0.20	0.02
Copper	327.4	Air/C_2H_2	0.001	0.02	2	0.15	0.02
Iron	248.3	Air/C_2H_2	0.006	0.03	3	0.16	0.02
Lead	217.0	Air/C_2H_2	0.01	0.06	6	0.20	0.02
Lithium	670.8	Air/C_2H_2	0.0008	0.025	2.5	–	–
Magnesium	285.2	Air/C_2H_2	0.0002	0.003	0.3	0.004	0.0004
Manganese	279.5	Air/C_2H_2	0.001	0.008	0.8	0.038	0.004
Molybdenum	313.3	N_2O/C_2H_2	0.015	0.2	20	–	–
Nickel	232.0	Air/C_2H_2	0.005	0.03	3	0.38	0.04
Potassium	766.5	Air/C_2H_2	0.0008	0.005	0.5	–	–
Selenium	196.0	N_2O/C_2H_2	0.08	0.5	5	0.23	0.02
Silver	328.1	Air/C_2H_2	0.001	0.02	2	0.02	0.002
Sodium	589.0	Air/C_2H_2	0.0002	0.002	0.2	–	–
Strontium	460.7	N_2O/C_2H_2	0.002	0.04	4	–	–
Thallium	276.8	Air/C_2H_2	0.01	0.02	2	0.25	0.03
Tin	286.3	N_2O/C_2H_2	0.12	0.6	60	0.50	0.05
Vanadium	318.5	N_2O/C_2H_2	0.02	0.4	40	0.26	0.03
Zinc	213.9	Air/C_2H_2	0.0006	0.005	0.5	–	–

[a]The sensitivity is the typical concentration of the element which produces a signal of 0.0044 absorbance.

[b]The sensitivity with respect to the soil sample is given assuming that a test sample of 1 g is digested and diluted to 100 ml. It is given as mg/kg dry matter.

4.3.3 Graphite furnace atomic absorption spectrometry

In graphite furnace atomic absorption spectrometry (GFAAS) or electrothermal atomic absorption spectrometry (ETAAS), an electrically heated graphite tube replaces the flame. A small volume of the acid digested sample (10–100 µl) is introduced directly into the tube, which is then heated in a programmed series of steps to remove the solvent and major matrix components, and then to atomise the remaining sample. All of the added analyte is atomised in a single pulse, and the atoms are retained within the tube (and the light path, which passes through the tube) for a significant period. As a result, sensitivity and detection limits are significantly improved but in cases of acid digestion of complex matrices, interferences and background effects are increased.

Electrothermal atomisation analysis times are longer than those for flame, and fewer elements can be determined using graphite furnace, mainly in the case of polluted soil samples. However, the enhanced sensitivity of this technique and its ability to analyse very small samples significantly expands the capabilities of atomic absorption. Graphite furnace atomic absorption spectrometry (GFAAS) applications are well documented, though not as completely as with flame. It has exceptional detection limit capabilities but with a limited analytical working range, no more than two orders of magnitude. This working range is a limited factor when analysing polluted soil sample due to the large dilution to be carried out, which may introduce an error in the final value. Sample throughput is significantly lower than other atomic spectroscopic techniques. Higher operator skill requirements are necessary than for FAAS.

To summarise, GFAA is a mature and reliable technique. But compared to flame, this technique needs great attention to detail and analyst skill to obtain good data. It is worth stressing that chemical interferences and background absorption will present problems in the determination of most elements in digests. The matrix interferences from the sample matrix such as salts and especially chlorides, which interfere with many elements (e.g. Cd, Cu, Pb), are the most difficult ones and must be handled with the use of an appropriate calibration (Slavin et al. 1984). The use of an appropriate background correction, especially below 350 nm, is mandatory. Deuterium background correction is commonly used but cannot compensate for structured background interference. When these interferences cannot be compensated for (e.g. for elements such as Se and As), the use of Zeeman effect which allows the atomic absorption and background absorbances to be measured at the same wavelength should lead to the correct results.

The importance of using an appropriate matrix modifier for the determination of each element in the sample digest must be highlighted and so also the use of analytical tools such as pyro-coated graphite tubes with a platform, maximum power and gas stop modes. Invariably, palladium salts figure prominently in any cocktail selected and the combination of $Pd+Mg$ (as NO_3 salts) is considered as a universal modifier. Diluted nitric acid, usually 1% v/v, is the preferred acid media for most analytical determinations by graphite furnace. A higher concentration of acids will increase the deterioration of graphite tubes, but up to 10% of mineral acid is not an analytical problem. If another acid in addition to nitric acid is required, a minimum amount should be used. This applies particularly to hydrochloric acid and, to a lesser extent, to sulfuric and phosphoric acids. Further information can be found in Hoenig and Guns (1995).

Only a few standard methods are available for the determination of trace elements in soil samples (ISO Standard 11047 and SW-846 EPA 7000A methods) and none specifically designed for polluted soil samples. Excellent sensitivity makes graphite furnace a possible technique for the determination of several commonly analysed elements at lower concentration in soil samples such as Pb, Cd, Cr, Ni, As, Se and Tl. In a graphite furnace technique, the sensitivity is expressed in peak height (P/H) or in peak area (P/A): that weight (expressed in picograms) of an element, which would yield an absorbance equal to 0.0044, also called characteristic mass. In Table 4.3, a comparison between instrumental performances obtained by FAAS and GFAAS is given. The main drawback of GFAAS is the slow speed of analysis and the limited number of elements that are suitable (e.g. for copper and zinc, the high sensitivity of GFAAS would require very large sample dilutions). For these reasons, very few large throughput contract analysis laboratories routinely use the GFAAS technique for polluted soil analysis.

4.3.4 Hydride generation atomic absorption and fluorescence methods

Hydride generation (HG) technique offers sensitivities greatly superior to those obtainable with other conventional methods except by ICP-MS and were first introduced by Holak (1969). They have been applied to the determination of hydride forming elements such as As, Bi, Ge, Sb, Se and Te by conversion to their hydrides by sodium borohydride reagent and transport into a suitable atomising device in an atomic absorption spectrometer. These vapour generation methods also have the advantage that chemical interference effects are small and well known (e.g. interference effects from other reducible metals such as Cu) because of the phase separation between the sample solution and the generated hydride vapour and gains in sensitivity of up to 50-fold compared with conventional AAS techniques. To perform hydride generation, procedures based on batch and continuous-flow have been used; but more recently, the use of flow injection analysis (FIAS) has been introduced. These systems offer an important alternative to batch methods because they are rapid, simple, versatile, very easy to automate and only a relatively small sample amount is required. The use of sodium borohydride rather than tin(II) chloride as both a chemical reductant and a hydrogen source is preferred in soil extract analysis. With this reagent, the hydrides are generated rapidly and completely. The conversion rate of the element to its hydride is reproducible and in addition interference effects due to high concentrations of Co, Cu, Fe, Hg or Ni are minimised. However, the efficiency of hydride formation depends on the oxidation state of the analyte. The sensitivity of arsenic(V) is about 70–80% of the sensitivity of As(III). For Sb, the higher oxidation state, Sb(V), has about 50% of the sensitivity of the lower state, Sb(III). In order to obtain the full sensitivity for As ad Sb, a mixture of potassium iodide and ascorbic acid is used to perform a pre-reduction step to the trivalent state. In case of selenium and tellurium, the difference between the oxidation states, (IV) and (VI), is very pronounced. The hexavalent state does not give any measurable signal, and a reduction step by boiling the sample in presence of hydrochloric acid is always required. In case of Ge and Sn, the oxidation state has no influence on hydride generation. Bismuth exists virtually only in the trivalent oxidation state, so that no pre-reduction step is necessary.

Hydride-forming elements (As, Sb, Se and Te) must be reduced prior to the measurement, which may be very difficult or impossible in digests containing high concentrations of oxidising acids such as nitric acid or perchloric acid. Transition metals may also interfere in ionic form in solution at high concentrations. Interferences caused by transition metals can be relatively easily eliminated or reduced by increasing the hydrochloric acid concentration, for selenium determinations, stronger acid concentration (50% v/v) and heating are required to complete the reduction of Se(VI) to Se(IV).

The use of atomic fluorescence rather than atomic absorption as a detector offers extremely good advantages in terms of linearity, specificity and sensitivity (Corns & Stockwell 1993). Nowadays, new instruments working with large volume autosamplers allow high sample throughput up to three samples/minute or 180 samples/hour in high throughput mode. ISO/TC/190 WG3 is now preparing a draft standard connected with the determination of As, Sb and Se in aqua regia soil extracts with ETAAS or hydride generation atomic spectrometry (ISO/CD 20280 (2001)).

To summarise, these techniques can be very useful for pushing down detection limits for hydride-forming elements mainly for unpolluted soil sample as shown in Table 4.4.

Table 4.4 Comparison between performance characteristics of hydride-forming elements for hydride generation with atomic absorption or fluorescence detection

Element	As	Bi	Sb	Se	Sn	Te
Sensitivity with HG-AAS in µg/L	0.20	0.20	0.15	0.30	0.40	0.15
Sensitivity with HG-AAS in mg/kg[a]	0.02	0.02	0.015	0.03	0.04	0.015
Limit of detection with HG-AFS in µg/L	0.01	0.01	0.01	0.02	–	0.01
Limit of detection with HG-AFS in mg/kg[a]	0.001	0.001	0.001	0.002	–	0.001

Sensitivity is the typical concentration of the element which produces a signal of 0.0044 absorbance.
Limits of detection are based on three times the standard deviation of the blank.
[a]The sensitivity with respect to the soil sample is given assuming that a test sample of 1 g is digested and diluted to 100 ml. It is given as mg/kg dry matter.

4.3.5   Cold vapour atomic absorption and fluorescence
methods for mercury

The most common technique for the determination of mercury in environmental samples is cold vapour atomic absorption spectrometry (CV-AAS) due to its simplicity and sensitivity. The flameless procedure was investigated by Hatch and Ott (1968) with a view to simplifying the apparatus required and improving the sensitivity of the method. The method is based on the unique properties of mercury. Elemental mercury has an appreciable vapour pressure at ambient temperature and the vapour is stable and monatomic. Mercury can easily be reduced to metal from its compounds. The mercury vapour may be introduced into a stream of an inert gas and measured by atomic absorption or atomic fluorescence of the cold vapour without the need of atomiser devices.

Inorganic mercury Hg(II) is usually reduced to metallic mercury by tin(II) chloride ($SnCl_2$), or sodium borohydride, then the reduced elemental mercury is liberated and swept to the absorption or fluorescence cell by the circulating inert gas.

Chemical interferences seldom occur in the cold vapour technique. Most acids and very many cations do not interfere even at high concentrations. Interference is caused by Ag, As, Bi, Cu, I, Sb and Se. The degree of interference is dependent on the reductant used. A complete suppression of the mercury signal in the presence of high concentration of Ag, Cu, Pd, Pt, Rh or Ru is noticed when using sodium borohydride as a reductant and is attributed to the reduction of these metals and subsequent amalgamation of mercury. The cold vapour method can be subjected to interferences from some volatile chloride and sulfur compounds. An adapted digestion technique is absolutely essential in order to determine the total mercury content if mercury is present as volatile methyl mercury species. The cold vapour technique may be fully automated by using a flow injection analysis (FIA) system, high sensitivity mercury analyser or atomic fluorescence detector. These systems offer high performance in terms of trace and ultra trace levels, reproducibility, excellent short- and long-term precisions, and high sample throughput. A comparison between atomic absorption and atomic fluorescence detection after acid digestion, including recommended and microwave-assisted digestions, on polluted soil samples was carried by Thomas and Morin (2000). They underlined that no significant differences can be found using these two methods of determination so that following this study, the ISO/TC/190 WG1 has prepared a draft new standard (ISO/DIS 16772 (2001)) connected with the determination of mercury in aqua regia soil extracts with cold-vapour atomic or cold-vapour atomic fluorescence spectrometry. An example of analytical performance for both atomic absorption and atomic fluorescence is given in Table 4.5.

Table 4.5 Performance characteristics for mercury by cold vapour techniques: obtained with the M6000A for CV-AAS and the Millennium PSA for CV-AFS

Element	Hg by CV-AAS	Hg by CV-AFS
Limit of detection in ng/L of Hg	8	6
Limit of detection in µg/kg[a]	0.8	0.6

Limits of detection are based on three times the standard deviation of the blank.
[a]The sensitivity with respect to the soil sample is given assuming that a test sample of 1 g is digested and diluted to 100 ml. It is given as mg/kg dry matter.
The detection limit for mercury is sometimes limited by background laboratory contamination rather than the actual detection method.

4.4 What is atomic emission spectrometry?

Atomic emission spectrometry is a process in which the light emitted by atoms or ions is measured. The emission occurs when sufficient thermal or electrical energy is available to excite a free atom or ion to an unstable energy state. Light is emitted when the atom or ion returns to a more stable configuration or ground state. The wavelengths of light emitted are specific to the elements present in the sample.

With the exception of better optical resolution needed, the basic instrument used for atomic emission is very similar to that used for atomic absorption with the difference that no primary light source is used for atomic emission. One of the most critical components for this technique is the atomisation source because it must also provide sufficient energy to excite the atoms as well as atomise them. The earliest energy sources for excitation were simple flames, but these often lacked sufficient thermal energy to be truly effective sources. The development in 1963 and the introduction in 1970 of the first commercial inductively coupled plasma (ICP) as a source for atomic emission dramatically changed the use and the utility of emission spectroscopy (Thompson & Walsh 1983).

4.4.1 Inductively coupled plasma atomic emission spectrometry

The ICP source is an argon plasma maintained by the interaction between a radio frequency (RF) field and ionised argon gas. The source is reported to reach temperatures as high as 10 000 K, with the sample experiencing useful temperature between 5500 K and 8000 K. These temperatures allow complete atomisation of elements, minimising chemical interference effects. The plasma is formed by a tangential stream of argon gas flowing between two quartz tubes. Radio-frequency (RF) power is applied through the coil, and an oscillating magnetic field is formed. The plasma is created when the argon is made conductive by exposing it to an electrical discharge, which furnishes seed electrons and ions. Inside the induced magnetic field, these charged particles are forced to flow in a closed annular path. As they meet resistance to their flow, heating takes place and additional ionisation occurs, the process occurs instantaneously and the plasma expands to its full dimensions.

As viewed from the top, the plasma has a circular, *doughnut* shape. The sample is injected as an aerosol through the centre of the doughnut. This characteristic of the source confines the sample to a narrow region and provides an optically thin emission source and a chemically inert atmosphere. Normally, samples are introduced as a solution into the plasma and argon is used as a carrier gas for the sample introduction. The much higher temperatures of the plasma compared to flame make ICP-AES more effective in detecting lower concentrations of refractory elements such as Ta, W and Zr, and rare earth elements.

A number of different chemical and physical effects, which occur during the transport process of the vapour phase, may cause interferences in ICP-AES. Chemical interferences include molecular compound formation, ionisation effects and solute vapourisation effects; they are highly dependent on the matrix type. They limit formation of free atoms in the vapour phase. These interferences are minimal with the ICP source due to the very high gas temperature, long residence times and inert atmosphere. In general, with increased plasma power, chemical interferences become smaller. Physical interferences are associated with the sample nebulisation and transport processes. Changes in viscosity and surface tension can cause significant errors, especially in the case of polluted soil samples digestion due to the high dissolved solids content. If physical interferences are present, diluting the sample, by using internal standard and by optimising the spray chamber and nebuliser, which is a very critical point, can reduce them. Care must be taken to ensure that potential interference effects are dealt with, especially when dissolved solids exceed 0.2% w/v. In fact, using a good compromise between nebuliser and spray chamber there is no difficulty in running samples with dissolved solids up to 1% w/v. Further information of elemental matrix effects in ICP-AES can be found in Todoli *et al.* (2002).

Spectral interferences are observed in every emission source and are most important in ICP-AES because emission lines that might be expected to be weak or non-existent in other sources are quite intense. Direct line coincidence occurs when the dispersion device is not capable of separating the analyte line from the matrix line. This type of interference is of importance for ICP-AES since analysis after total or partial digestion of polluted soil samples often results in high levels of interfering elements such as Al, Ca, Fe, K, Na, Mg and Ti. However, with recent instruments offering high resolution or unlimited selection of analyte wavelengths, moving to another wavelength that does not exhibit the interference may solve these problems. Matrix matching the standard solution may solve some of the interference problems. Alternatively, different techniques for correction of this type of interferences are available with the software of the instrument used (e.g. spectral background correction, spectral background subtraction or multi-component spectral fitting). Also, when plasma is working in robust conditions, matrix effects resulting from a change in plasma can be minimised (Dubuisson et al. 1998). A validation of ICP-AES methods for soil and sediments for the determination of Cd, Co, Cr, Cu, Mn, Ni, Pb and Zn was carried out by Bettinelli et al. (2000a,b) after using aqua regia and total digestion procedures by using microwave assisted digestion. They highlighted that the precision data for elements present at lower concentrations (e.g. Pb, Cd) are slightly poorer for the total digestion procedure than those obtained for aqua regia solutions. This can reflect the variability associated with the analysis of more complex solutions derived from the complete dissolution of the sample.

It is important to appreciate that detection limits, sensitivity and optimum ranges of the elements will vary with the make and model of ICP-AES spectrometer. The spectrometer may be of the simultaneous (polychromator) or sequential (monochromator) type with air-path ($\lambda > 195$ nm), inert gas purged, or vacuum optic ($\lambda > 160$ nm) and also different types of detectors can be used (photomultiplier tube (PMT), segmented-array charge-coupled device (SCD), charge coupled device (CCD), charge injection device (CID)). In addition, the sensitivity and the detection limits of course depend on the radial or axial viewing system. Axial systems are approximately 5–10 times more sensitive, but subject to more interference effects.

When working under specific operating conditions (vacuum optic or inert gas purged) sulfur, phosphorus and bromine can be determined on digested samples. Because of so many differences between various models on the market, no detailed operating conditions or recommended wavelength can be given without giving the name and the make of the instrument. It is recommended that sensitivity, instrumental detection limit, precision, linear dynamic range and interference effects are established for each individual analyte line chosen on that particular instrument. All measurements must be within the instrument linear range where calibration factors are valid. The very wide working range is a feature of ICP analysis, and trace element (mg/kg) and major element (g/kg) concentrations can be determined on the same sample without the need for dilutions. Table 4.6 lists elements for which this technique applies, example of analytical wavelengths, and typical estimated instrument detection limits (IDL) using conventional pneumatic nebulisation and axial or radial viewing.

ICP-AES is the best overall multi-element atomic spectroscopy technique for soil metal analysis, with excellent sample throughput and very wide

Table 4.6 Suggested wavelengths and estimated detection limits for selected elements obtained using ICP-AES Varian Vista-MPX megapixel CCD detector features (by the permission of Varian Analytical Instruments)

Metal	Wavelength (nm)	Axial viewing		Radial viewing	
		Detection limit (μg/L)[a]	Detection limit (mg/kg)[b]	Detection limit (μg/L)[a]	Detection limit (mg/kg)[b]
Aluminium	396.068	0.9	0.09	4	0.4
Antimony	206.83	3	0.3	16	1.6
Arsenic	188.98	3	0.3	12	1.2
Arsenic	193.696	4	0.4	11	1.1
Barium	233.527	0.1	0.01	0.7	0.07
Barium	455.403	0.03	0.003	0.15	0.02
Beryllium	313.107	0.05	0.005	0.15	0.02
Cadmium	214.439	0.2	0.02	0.5	0.05
Calcium	396.847	0.01	0.001	0.3	0.03
Calcium	317.933	0.8	0.08	6.5	0.7
Chromium	267.716	0.5	0.05	1	0.1
Cobalt	238.892	0.4	0.04	1.2	0.1
Copper	327.395	0.9	0.09	1.5	0.1
Iron	238.204	0.3	0.03	0.9	0.09
Lead	220.35	1.5	0.15	8	0.8
Lithium	670.783	0.06	0.006	1	0.1
Magnesium	279.55	0.05	0.005	0.1	0.01
Magnesium	279.8	1.5	0.15	10	1
Manganese	257.61	0.1	0.01	0.13	0.01
Molybdenum	202.03	0.5	0.05	2	0.2
Nickel	231.6	0.7	0.07	2.1	0.2
Phosphorus	177.43	4	0.4	25	2.5
Potassium	766.491	0.3	0.03	4	0.4
Rubidium	780.03	1	0.1	5	0.5
Sulfur	181.962	4	0.4	13	1.3
Selenium	196.03	4	0.4	16	1.6
Silver	328.068	0.5	0.05	1	0.1
Sodium	589.59	0.2	0.02	1.5	0.2
Strontium	407.77	0.02	0.002	0.1	0.01
Thallium	190.79	2	0.2	13	1.3
Tin	189.93	2	0.2	8	0.8
Titanium	336.12	0.5	0.05	1	0.1
Vanadium	292.4	0.7	0.07	2	0.2
Zinc	213.86	0.2	0.02	0.8	0.08

[a]Typical 3-sigma detection limits using 30 seconds integration time.
[b]The detection limit with respect to the soil sample is given assuming that a test sample of 1 g is digested and diluted to 100 ml. It is given as mg/kg dry matter.

analytical range. Operator skill requirements are intermediate between flame and electrothermal atomic absorption spectrometry. Generally, use of this technique is restricted to operators who are knowledgeable in the correction of spectral, chemical and physical interferences, as described above.

4.5   What is inductively coupled plasma mass spectrometry?

The first ICP-MS instrument was introduced in the market in 1983 following the development of the technique by Gray (1974, 1975). A basic ICP-MS is the synergistic combination of inductively coupled plasma with a quadrupole mass spectrometer. ICP-MS uses the ability of the argon ICP to efficiently generate singly charged ions from the elemental species within the sample. These ions are then directed into a quadrupole mass spectrometer passing through a sophisticated interface. The function of the mass spectrometer is similar to that of the monochromator in an AAS or ICP-AES system. However, rather than separating light according to its wavelength, the mass spectrometer separates the ions introduced from the ICP according to their mass-to-charge ratio. Ions of the selected mass/charge are directed to a detector, which quantitates the number of ions present. Due to the similarity of the sample introduction and data handling techniques, using an ICP-MS is very much like using an ICP emission spectrometer. ICP-MS combines the multi-element capabilities and broad linear working range of ICP-AES with exceptional detection limits much better than graphite furnace AAS. One of the prime advantages of ICP-MS over most of the other techniques considered is that it allows the direct determination of element isotopic concentrations or isotopic ratios. This ability to discriminate between isotopes of elements can lead to interesting applications (e.g. study of elements as tracers in soil columns). Only high-resolution instruments allow determination of isotope ratios with the sort of precision required by environmental scientists, because of the low background achievable. This is a true multi-element technique at trace and ultra-trace concentration levels, the detection limits are nowadays at the 0.01–500 ng/l range in the final sample digest. Moreover, practically all elements can be determined and the detection limits do not differ much from one element to another when using basic ICP with quadrupole mass spectrometr. However, the detection limits of some light elements are higher than those of the heavier elements, which is due to isobaric interferences with polyatomic ions. However, most polyatomic ion interferences in ICP-MS are now well documented. ICP-MS interferences are a complex issue, with several factors contributing to the suppression or enhancement in analytical response observed, and will not be fully discussed in this chapter. We will describe briefly the various types of interferences. Further details can be found in Jarvis *et al.* (1992), Grenville-Holland and Tanner (2001) and ISO/DIS 17294-1 (2002).

Isotopes of different elements that form ions of the same nominal mass-to-charge ratio are not resolved by the quadrupole mass spectrometer, and cause isobaric elemental interferences. Typically, ICP-MS instrument operating software will have all known isobaric interferences entered, and will perform necessary correction calculations automatically. However, increasing the spectrometer resolution power from $m/\Delta m = 400$ to 1000 minimises abundance sensitivity interferences.

Polyatomic (background molecular and matrix molecular) ion interferences are caused by ions consisting of more than one atom and having the same nominal mass-to-charge ratio as the isotope of interest; it is important to appreciate that Ar, H, N and O are always present in the plasma. Most of the common molecular ion interferences have been identified and are listed in Table 4.7. Because of the severity of the chloride ion interference on important analytes, particularly arsenic, vanadium and selenium, hydrochloric acid or aqua regia is not recommended for use in the preparation of any samples to be analysed by quadrupole ICP-MS. Because chloride ion is present in most soil samples, it is critical to use chloride correction equations for affected masses. For example, when working in the presence of 5% of HCl, ^{51}V, ^{75}As and ^{77}Se masses will suffer interference from ^{35}ClO, Ar^{35}Cl and Ar^{37}Cl respectively. Regarding these interferences, software interference correction can be used to correct spectral interference due to chloride. Also, alternative options such as Collision Reaction Cell, Dynamic Reaction Cell, Collision Cell Technology and Octopole Reaction System may be used to solve specific interferences caused by polyatomic ions, but using these accessories the ICP-MS loses its multi-element capability. Polyatomic interferences are strongly influenced by instrument design and plasma operating conditions, and can be reduced in some cases by careful adjustment of nebuliser gas flow, spray chamber temperature and other instrument operating parameters. This is why nitric acid is commonly used in conjunction with ICP-MS analysis although many laboratories have developed aqua regia methods using appropriate correction techniques as mentioned above.

Physical interferences include differences in viscosity, surface tension and dissolved solids between sample and calibration standards. To minimise these effects, solid levels in analytical samples should not exceed 0.2% w/v, which requires the dilution of the digest before analysis. Use of several internal standards for correction of physical interferences is suitable when carrying out analysis of soil digest.

Due to the very high sensitivity of the ICP-MS technique, memory (carry over) effects may occur when analytes from a previous sample are measured in the current sample. In cases where analysis of highly polluted soil digests is carried out, memory effects can occur, they may be indications of problems in the sample introduction system. Severe memory interferences may require disassembly and the cleaning of the entire sample introduction system, including the plasma torch and the sampler and skimmer cones. Due to these memory

Table 4.7 Recommended isotopes for selected elements and blank equivalent concentrations (BEC) for selected elements obtained using ICP-MS X7 CCT Thermo-Elemental (by the permission of Thermo-Elemental)

Metal	Recommended isotopes[a]	Direct aspiration			
		Measured isotope[b]	BEC (ng/L)	BEC (µg/kg)[c]	Plasma mode[d]
Aluminium	*27*	27	1.04	0.1	Standard
Antimony	121, 123	121	0.1	0.01	Standard
Arsenic	*75*	75	0.04	0.05	Standard
Barium	134, **135**, 136, 137, 138	138	0.05	0.005	Standard
Beryllium	*9*	9	0.19	0.02	Standard
Bismuth	*209*	209	0.01	0.001	Standard
Cadmium	106, 110, **111**, 112, 113, *114*, 116	114	0.20	0.02	Standard
Calcium	*40*, 42, 43, **44**, 46, 48	40	2.65	0.27	CCT
Chromium	**50**, *52*, **53**, 54	52	534	54	Standard
Cobalt	*59*	59	0.15	0.02	Cool
Copper	*63*, **65**	63	0.72	0.07	Standard
Holmium	*165*	165	0.04	0.004	Standard
Indium	*115*, 113,	115	0.03	0.003	Standard
Iron	**54**, *56*, 57, 58	56	1.03	0.1	Cool
Lanthanum	*139*	139	0.04	0.004	Standard
Lead	204, **206**, **207**, **208**	208	0.03	0.003	Standard
Lithium	6, *7*	7	0.07	0.01	Cool
Magnesium	*24*, **25**, **26**	24	0.14	0.02	Cool
Manganese	*55*	55	15	1.5	Standard
Mercury	199, **200**, 201, *202*	202	0.30	0.03	Standard
Molybdenum	92, 94, 96, **97**, *98*	98	0.52	0.05	Standard
Nickel	*58*, **60**, **61**, 62, 64	58	4.50	0.5	Cool
Potassium	*39*	39	1.38	0.1	Cool
Rhodium	*103*	103	0.05	0.005	Standard
Scandium	*45*	45	190	19	Standard
Selenium	74, **76**, **77**, *78*, 80, **82**	80	0.84	0.09	CCT
Silver	*107*, **109**	107	0.08	0.008	Standard
Sodium	*23*	23	0.18	0.02	Cool
Strontium	84, 86, 87, *88*	88	0.07	0.007	Standard
Terbium	*159*	159	0.04	0.005	Standard
Thallium	203, **205**	205	0.01	0.001	Standard
Tin	**118**, *120*	120	0.74	0.08	Standard
Vanadium	**50**, *51*	51	3.00	0.3	Standard
Ytttrium	*89*	89	0.04	0.05	Standard
Zinc	*64*, **66**, **67**, **68**, 70	64	1.20	0.1	Standard

[a] Recommended isotopes to be monitored according to EPA-Methods 6020A 1998. The most generally useful isotopes are in boldface and the most abundant is italicised.

[b] Mass used to obtain the blank equivalent concentration (BEC), see explanation in the text.

[c] The BEC with respect to the soil sample is given assuming that a test sample of 1 g is digested and diluted to 100 ml. It is given as mg/kg dry matter.

[d] Plasma operating conditions (standard, cool plasma or collision cell technology (CCT)).

effects, analysis of mercury is very difficult to carry out by ICP-MS technique unless the gold stabilisation technique is used (Fatemian 1999a,b).

Ionisation interferences result when moderate (0.1–1% m/m) amounts of a matrix ion change the analyte signal. This effect, which usually reduces the analyte signal, also is known as *suppression*. The use of an adapted internal standard is suitable to correct for this type of suppression.

The sample throughput for ICP-MS is typically 20–30 element determinations per hour depending on such factors as the concentration levels and required precision. One of the main features of ICP-MS is the near uniform sensitivity across the mass range. This feature would allow standardless semi-quantitative analysis for screening purposes and therefore the rapid estimation of trace metal contents in soil samples.

In spite of all the advantages of ICP-MS in terms of sensitivity and selectivity, the need, in most laboratories, for sample dissolution remains a problem when performing measurements by this technique. Unfortunately, multi-elemental analysis using ICP-MS for a large number of elements in soil samples of different origin or different levels of contamination can be very time consuming unless the various sample concentrations are taken into account, especially in cases of investigation of sites where the degree of pollution is unknown. Recent instruments have increased the dynamic range to allow measurement of ng/l to mg/l using the following set-up:

First, a special interface design, with specific set of cones, allows the measurement of over 200 mg/l of Na with same ng/l detection limits for other trace elements higher in the mass range. Second, in the experimental method it is also possible to select higher resolution mode ($m/\Delta m = 1000$) on the quadrupole for each major analyte with normal resolution mode ($m/\Delta m = 400$) for other trace analytes. This allows the measurement of up to 1 g/l Na when used in conjunction with the special interface. A validation of ICP-MS methods for soil and sediment for the determination of trace metals has been performed by Bettinelli *et al.* (2000a,b) after aqua regia and total digestion procedures by using microwave-assisted digestion. The precision with these two methods is, on average, lower than 10% (RSD).

This method is ideally used by analysts experienced in the use of ICP techniques, the interpretation of spectral and matrix interference, and procedures for their correction. It is preferable to demonstrate analyst proficiency through analysis of performance evaluation samples (e.g. reference or certified reference materials) before the generation of routine analysis data. Use of ICP-MS method should be restricted to competent analysts who are knowledgeable in the recognition and in the correction of spectral, chemical and physical interferences.

Table 4.7 presents typical recommended analytical masses and indicative blank equivalent concentration (BEC) for analytes. The quantification limits

(LOQ) will typically be higher than the instrumental detection limit (IDL), because of background analyte and matrix-based interferences. The BEC (blank equivalent concentration) used in Table 4.7 is the apparent concentration of an analyte normally derived from intercepted point of its calibration curve or by reference of the actual counts for that analyte in a blank solution. The BEC gives a good indication of the blank level, which will affect the IDL. Most often, the detection limits are calculated as three times the normal standard deviation of the BEC in a within batch replicate analytical measurement of a blank solution. Therefore, if the instrument is stable enough, this will give a better IDL than the BEC itself. The BEC is a combination of the contamination of the analyte in the solution, the residual amount of the analyte in the spectrometer and the contribution of any polyatomic species in the analyte mass.

4.6   How to select the proper technique?

The purpose of many chemical analyses is to assess compliance with limit values specified in a regulation. Compliance may require the amount of a specified measurand to be above or below a single limit value or within a given range. In some cases the result, including its uncertainty interval, will be sufficiently different from the limit value for the state of compliance (pass/fail) to be unambiguous. In other cases the state of compliance will be unclear (i.e. the compliance value falls within the range of the determined analyte value plus or minus its associated uncertainty (see also Figure 1.3 of Chapter 1)). With the availability of a variety of atomic spectroscopy techniques such as flame, electrothermal atomisation, inductively coupled plasma emission and ICP mass spectrometry, analysts must decide which technique is the best suited for the analytical problems of their laboratory. Because atomic spectroscopy techniques complement each other so well, it may not always be clear which technique is optimum.

A clear understanding of the analytical problem in the laboratory and the capabilities provided by the different techniques is necessary. Important criteria for selecting an analytical technique include detection limits, analytical working range, sample throughput, cost, interferences, ease of use and the availability of proven methodology, and also whether recommended standard methods are available. In making analytical selections, decision makers or site owners should carefully consider the detection limits of the various techniques. A detection limit has an uncertainty of $\pm100\%$; in other words, a detection limit of 1 mg/kg implies an uncertainty of ~1 mg/kg. When a customer requires very precise data in the region of one to five times the detection limit of a particular technique, then it is recommended that an alternative technique is selected, which offers a lower detection limit. For example, if precise data in the range of 1–5 mg/kg for toxic metals are needed, then the ICP-MS technique is recommended. This offers detection limits in the range of 0.01–0.5 mg/kg for these metals. As a general rule, the detection limit should be ten times lower than the compliance or limit value. In case of site investigation or detection of pollution of soils, the main selection criteria for atomic spectroscopic techniques are concentration range and analytical throughput. Where the selection is based on analyte detection limits, flame and ICP-AES are favoured for moderate to high levels (e.g. 10–1000 mg/kg), while GFAAS and ICP-MS are favoured at lower levels. Figures 4.3 and 4.4 show, respectively, typical dynamic ranges

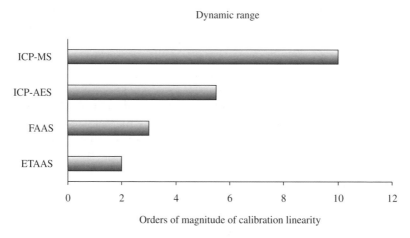

Figure 4.3 Typical detection limit ranges for the major atomic spectroscopic techniques.

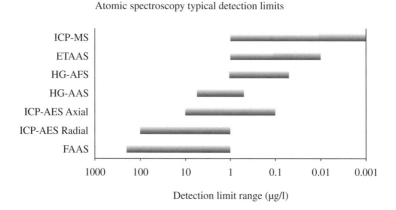

Figure 4.4 Analytical working ranges for the major atomic spectroscopic techniques.

with a single set of instrument conditions and typical instrumental detection limit ranges for these techniques.

Advances in modern analytical instrumentation have made it easier to determine many elements during a single analysis, using ICP-AES and ICP-MS, which are multi-element techniques. These techniques are favoured where large number of samples are to be analysed, and allow semi-quantitative analysis for screening purposes and therefore the rapid estimation of the trace metal contents in soil samples from a site investigation. When an instrument is capable of performing simultaneous multi-element analysis, it is generally as expensive for the laboratory to measure one element, as it is to measure several. The various ICP-MS packages offer an attractive mix of elements and detection limits but these packages are applicable only to slightly polluted soils because the electron multiplier detectors are easily saturated by even lightly mineralised material. It goes without saying that detection limits, sensitivity and optimum ranges of metals will vary with the type of matrices and model of instruments. The instrumental performance data shown in Tables 4.3–4.7 only provide some indication of the detection limits achievable by these various techniques. Therefore, it is the responsibility of the analyst to demonstrate the accuracy and precision of the method in the type of samples to be analysed. The analyst is always required to monitor potential sources of interferences and take appropriate action to ensure data of known quality under ISO 17025 (see also Chapter 1).

4.7 Overview of trace element speciation

This section is not intended to be a full review on speciation; there are many papers and books dealing with the subject e.g. Quevauviller *et al.* 1995; Quevauviller 1998. In this chapter, the determination of redox species, mainly chromium and organometallic compounds, will be briefly addressed. In principle, the difference between the two types of determination is negligible since the species determination is mostly preceded by an extraction.

The environmental impact of metallic contaminants in soils is dependent both on the chemical speciation of the metal and the response of the matrix to biological and physicochemical conditions. These factors are responsible for the mobilisation of the metal from the solid into the aqueous phase hence transport within the immediate vicinity, has an impact on the rate of dispersal, dilution, uptake and transfer into living systems. The impact of changing environmental conditions on the contaminant inventory can be to enhance or moderate these phenomena, with subsequent consequences for the broader risk assessment of the contaminants. During the last ten years, extensive research in analytical chemistry was initiated to develop highly specific and sensitive methods for measuring potentially harmful substances in various matrices but only a few were developed in polluted soil samples. The International Union for Pure and Applied Chemistry (IUPAC) recommendation for the definition of terms related to chemical speciation of elements is adopted here.

Chemical species (Chemical element). Specific form of an element defined based on its isotopic composition, electronic or oxidation state, and/or complex or molecular structure.

Speciation analysis (Analytical chemistry). Analytical activities to identify and/or measure the quantities of one or more individual chemical species in a sample.

Speciation of an element; speciation. Distribution of an element amongst defined chemical species in a system. The term fractionation has been defined as follows.

Fractionation. Process of classification of an analyte or a group of analytes from a certain sample according to physical (e.g. size, solubility) or chemical (e.g. bonding, reactivity) properties. The terms are described in detail in the original manuscript (Templeton *et al.* 2000). See also Chapter 9.

In the scientific literature, speciation studies were mainly carried out by using leaching of agricultural soils with variety of differing (relatively mild) extraction reagents. The determination of essential elements and potentially toxic trace elements by selective extraction procedures (fractionation) has been used successfully for decades for the diagnosis of plant growth disturbances and diseases of animals, and as a means for avoiding them. The goal of these investigations is to determine the bioavailability of contaminants in agricultural soils rather than their total concentration. Although the majority of these procedures were determined empirically they nevertheless give good correlation for numerous elements. Kersten and Förstner (1995) review the various extraction procedures.

Since, with the sequential extraction procedure of soils with a variety of differing reagents only relatively small element concentrations are leached out, especially with weak extracting agents, FAAS is sufficiently sensitive only for a few elements such as Zn, Fe, Mn and Cu. The ICP-AES technique is much better in terms of sensitivity and range of elements to be analysed. However, when low levels of trace elements are expected, it is necessary to use a graphite furnace or, even better, the ICP-MS technique. However, it is important to be aware of extracting agent matrix interferences. Many of these will be present at high concentrations. Thus all methods must be carefully validated. The reproducibility and comparability of selective sequential leaching is frequently problematic, and the correct assignment of the various extracts to the corresponding species is not always unambiguous since redistribution can take place during the process. In more recent times improved control of parameters such as extraction temperature and time, or the application of the microwave-assisted system, has led to an improvement in reproducibility. In addition, certified reference materials are now available for the extractable portion of agricultural soils, but unfortunately there is nothing relating to polluted soils.

The non-chromatographic procedure for speciation analysis includes mainly extraction of Cr(VI) in weak alkaline medium as proposed in the USEPA method (SW-846 Method 3060A (1998)). This method is of significance due to the differences in toxicity between Cr(III) and Cr(VI). Cr(VI) is a human carcinogen and Cr(III) is an essential dietary component for humans (Herold & Fitzgerald 1994). Usually, once Cr(VI) is solubilised, the digest is analysed after it reacts with specific reagents to form a chromophoric product, and its absorbance is measured spectrophotometrically at 540 nm. An alternative determination technique is possible by using ion chromatography with a post-column reaction. This method is easy to use and could be carried out for investigation on chromium-polluted sites. The major problem is that results are fully dependent on the procedure. Data are not consistent with other extraction procedures such as the DIN 19734 (1990) method.

Regarding other elements such as tin, selenium, arsenic and mercury, the determination of elemental species requires analytical technology with sufficient selectivity and sensitivity to resolve and quantitate the individual species at trace levels, since an individual toxic species may constitute only a very small fraction of an element's total concentration in a sample. In order to obtain satisfactory answers to the questions of trace metal speciation, a sensitive and selective technique is needed. The growing interest in speciation has been generated by significant improvements in technology and by the growth of speciation in the monitoring and regulation of trace metals. One of the promising technologies for meeting these requirements is high performance liquid chromatography (HPLC) linked to ICP-MS because of its detection limits of sub ng concentrations. An example of HPLC-ICP-MS system is shown on Figure 4.5.

This technique is especially useful in carrying out automated, high throughput speciation analysis. The multi-elemental capability of ICP-MS can be used for simultaneous speciation analysis of As, Se, Sb and Tl (Casiot *et al.* 1999). It is possible to highlight much work focused on sediments or sludges probably due to the regulation on disposal and seawater quality. For example, organotin compounds have been widely studied in solid waste using many analytical methods, the most cited analytical technique being gas chromatography. This technique can be interfaced with atomic emission (AED), flame photometry (FPD), pulse flame photometry (PFPD), mass spectrometry (MS), atomic absorption spectrometry (AAS) and more recently with inductively coupled plasma-mass spectrometry (ICP-MS). There are problems when trying to interface liquid chromatography (LC) with ICP-MS, as many organic solvents have a high vapour pressure and nebulisation of these solvents in the ICP-MS spray chamber can extinguish the plasma and carbon build up on the cone; the latter effect can be overcome by adding a flow of oxygen to the argon flow in conjunction with platinum cones. This complex technique is not often used in routine laboratories.

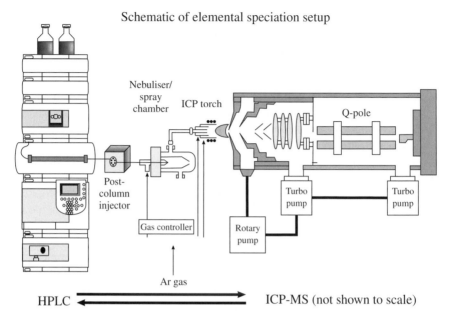

Figure 4.5 Principle of high performance liquid chromatography (HPLC) linked to inductively coupled plasma mass spectrometry (ICP-MS). (Adapted from: Fairman & Walhen 2001.)

In some recent ICP-MS systems the spray chamber is Peltier cooled to reduce solvent loading on the plasma. Further information about organotin speciation can be found in (Drickx *et al.* 1995) and for LC techniques in (Ebdon *et al.* 1995).

A selenium speciation study was carried out by Séby *et al.* (1997) on a seleniferous soil using hydride generation quartz furnace atomic absorption spectrometry (HG-QFAAS) method and after alkaline extraction. The speciation was performed in order to identify and determine inorganic and organic selenium forms.

To date, little work has been cited in the literature with respect to arsenic speciation of polluted soil. A feasibility study on the identification and monitoring of arsenic species in polluted soil and sediment samples (Thomas *et al.* 1997) has been reported. In this study, polluted soil samples were extracted in phosphoric acid media using an open vessel microwave-assisted extraction system. The determination of arsenic species was investigated using an on-line system involving HPLC-ICP-MS system. The speciation was performed to identify As(III), As(V) and monomethylarsonic acid (MMA) and dimethylarsinic acid (DMA). The proposed method had the potential to form the basis of a routine procedure for monitoring the behaviour of arsenic species in soils. This extraction procedure was recently applied to contaminated soils, by Garcia-Manyes *et al.* (2002), after modification by addition of ascorbic acid to phosphoric acid; detection of previous arsenic species was carried out with the coupling liquid chromatography (LC)-ultraviolet (UV) irradiation-hydride generation (HG)-ICP-MS or AFS detection.

Monomethylmercury (MeHg) is of particular concern due to its extreme toxicity and its ability to bioaccumulate in fish tissues; this compound is routinely determined in fish samples to control the contamination levels prior to market sales. More recently, monitoring of MeHg contents in sediments has been started in research laboratories for the purpose of pollution monitoring and geochemical studies (e.g. studies of methylation/demethylation processes), but nothing has been done specifically on contaminated or polluted sites. However, much of the work reported on speciation analysis has been performed using methods that have not been properly validated. Indeed, the techniques developed for the determination of element species in solid matrices involve, in most cases, a succession of analytical steps (extraction, derivatisation, separation, detection), which may all be prone to systematic errors. Over ten years ago, when these methods were developed, they were poorly validated due to a lack of evaluation programmes and suitable certified reference materials.

The situation has improved recently, thanks to the development of more sensitive and specific analytical techniques, the organisation of interlaboratory studies and the availability of certified reference materials (CRMs), produced by a number of producers such as the European Commission through the measurement & testing programme (SMT); the National Research Council of Canada (NRCC); the International Atomic Energy Agency (IAEA); the National Institute of Standards and Technology (NIST), USA and the National Institute of Environmental Studies (NIES), Japan and also Specialist Techniques Group at Laboratory of the Government Chemist (LGC), Teddington, UK. Initially, most of the reference materials were of biological origin but recently, sediment reference materials have been certified and currently a feasibility study is on the way to produce two polluted soil samples for arsenic speciation (Ebdon 2000).

To conclude, it can be said that speciation techniques are being improved due to the growing importance of speciation in the monitoring of trace metals. There is considerable pressure for speciation methods to be used in routine analysis, particularly with regard to legislative demands. However, this is only likely to be achieved with the development of robust validated methods utilising relevant soil reference materials in order to obtain reliable data.

References

Bettinelli, M., Baffi, C., Beone, G.M. & Spezia, S. (2000a) *Atomic Spectroscopy*, **21**, 50–59.
Bettinelli, M., Baffi, C., Beone, G.M. & Spezia, S. (2000b) *Atomic Spectroscopy*, **21**, 60–70.
Casiot, C., Donard, O.F.X. & Potin-Gautier, M. (1999) *Spectra Analyse*, **206**, 17–22.
Corns, W.T. & Stockwell, P.B. (1993) *J. Anal. At. Spectrom.*, **8**, 71–77.

Cresser, M.S. (1994) *Flame Spectrometry in Environmental Chemical Analysis – A Practical Guide*, N. Barnett (eds), Deakin University, Victoria, Australia, The Royal Society of Chemistry (RSC). ISBN 0–85186–734–0.

Danish Standard (1982) Determination of metals in water, sludge and sediments – general guidelines for determination by atomic absorption spectrometry in flame. Ds 259.

Danish Standard (1990) Determination of metals in water, sludge and sediments – determined by flameless for determination by atomic absorption spectrometry – electrothermal atomisation in graphite furnace – General principles and guidelines. Ds 2210.

DIN 19734 (1999) German standard method: Soil Quality – Determination of chromium(VI) of water, wastewater and sludge. Sludge and sediments. Digestion using aqua regia for subsequent determination of acid-soluble portion of metals. Deutsche Norm, DIN 38414, Part 7.

Drickx, W.M.R., Lobinski, R. & Adams, F.C. (1995) *Quality Assurance for Enviromental Analysis*, Quevauviller/Mayer/Griepink (eds), **17**, 357–409. Elsevier Science, ISBN 0–444–89955–3.

Dubuisson, C., Poussel, E., Mermet, J.M. & Tolodi, J.L. (1998) *J. Anal. At. Spectrom.*, **13**, 63–67.

Ebdon, L. (2000) Feasibility studies for speciated CRMs for arsenic in chicken, rice, fish and soil and selenium in yeast and cereal, G6RD-CT-2000-00473, SM&T program.

Ebdon, L., Hill, S.J. & Rivas, C. (1995) *Trends in Analytical Chemistry*, **17**(5), 277–288.

Ebdon, L., Pitts, L., Cornelis, R., Crews, H., Donard, O.F.X. & Quevauviller, P. (eds) (2001) Trace Element Speciation for Environment, Food and Health, The Royal Society of Chemistry (RSC). ISBN 0–85404–459–0.

European Standard (1999) Characterisation of waste – Microwave assisted digestion with hydrofluoric (HF), nitric (HNO$_3$) and hydrochloric (HCl) acid mixture for subsequent determination of elements in waste, prEN 13356, Brussels.

European Standard (1999) Characterisation of waste – Digestion for subsequent determination of aqua regia soluble portion of elements in waste, prEN 13357, Brussels.

European Standard (2000) Characterisation of sludges: Aqua regia extraction methods – Determination of trace elements and phosphorus, EN 13346.

Fairman, B. & Wahlen, R. (2001) *Spectrosopy Europe*, **13**(5), 16–22.

Fatemian, F., Allibone, J. & Walker, P. (1999a) *J. Anal. At. Spectrom.*, **14**, 235–239.

Fatemian, F., Allibone, J. & Walker, P. (1999b) *Analyst*, **124**, 1233–1236.

Garcia-Manyes, S., Jimenez, G., Padro, A., Rubio, R. & Rauret, G. (2002) *Talanta*, **58**, 97–109.

Gray, A.L. (1974) *Proc. Soc. Anal. Chem.*, **11**, 182–183.

Gray, A.L. (1975) *Analyst*, **100**, 289–299.

Gray, A.L. (1993) *Application of Plasma Source Mass Spectrometry II*, Grenville-Holland & Eaton, A.N. (eds), (1992) (Special Publication, No. 124), The Royal Society of Chemistry (RSC). ISBN 0–85186–465–1.

Grenville-Holland & Tanner, S.D. (2001) *Plasma Source Mass Spectrometry: The New Millennium* (Special Publications, No. 267), The Royal Society of Chemistry (RSC). ISBN 0–85404–895–2.

Grosser, Z.A. & Schneider, C.A. (1996) *Atomic Spectroscopy*, **17**, 209–210.

Harrison, R.M. & Rapsomanikis, S. (eds) (1989) *Environmental analysis using chromatography interfaced with atomic spectroscopy*, Ellis Horwood Ltd, Chichester.

Hatch, W.R. & Ott, W.L. (1968) *Anal. Chem*, **40**(14), 2085.

Herold, D.A. & Fitzgerald, R.L. (1994) In *Handbook on metals in clinical and analytical chemistry*, Hans G. Seiler, Astrid Sigel and Helmut Sigel (eds), New York, pp. 321–332. ISBN 0–8247–9094–4.

Hester, R.E. & Harrison, R.M. (2001) *Assessment and Reclamation of Contaminated Land*, Hester and Harrison (eds), **16**, Royal Society of Chemistry (RSC). ISBN 0–85404–275–X.

Hjelmar, O. (1999) Determination of total or partial trace element in soil and inorganic waste material, NT Technical report 446, 44p.

Hoenig, M. (1998) *Trends in Analytical Chemistry*, **17**(5), 272–276.

Hoenig, M. & Guns, M.F. (1995) *Quality Assurance for Enviromental Analysis*, Quevauviller/Mayer/Griepink (eds), **17**, 357–409. Elsevier Science, ISBN 0–444–89955–3.

Hoenig, M. & De Kersabiec, A.M. (1996) *Spectrochim Acta*, B51, 1297.

Hoenig, M. & Thomas, P. (2002) *Techniques de l'ingénieur*, P4150, **P5**, 8.

Holak, W. (1969) *Anal. Chem*, **41**, 1712.

ISO (1995) Soil Quality – *Extraction of trace elements soluble in aqua regia*. International standard ISO 11466.

ISO 11047:1998 Soil Quality – Determination of cadmium, chromium, cobalt, copper, lead, manganese, nickel and zinc – Flame and electrothermal atomic absorption spectrometric methods.

ISO (2001) Soil Quality – *Determination of total trace element content. Part 1: Digestion with hydrochloric and perchloric acids*, International standard ISO 14869.

ISO (2001) Soil Quality – *Determination of mercury in aqua regia soil extracts with cold-vapour atomic spectrometry or cold vapour atomic fluorescence spectrometry*, ISO/DIS 16772.

ISO (2001) Soil Quality – *Determination of arsenic, antimony and selenium in aqua regia soil extracts with electrothermal or hydride generation atomic absorption spectrometry*, draft ISO/CD 20280.

ISO/DIS 17294, Water Quality – Application of inductively coupled plasma mass spectrometry (ICP-MS) – Part 1: General guideline; Part 2: Determination of 61 elements (in preparation).

Jarvis, K.E., Gray, A.L. & Houk, R.S. (1992) *Handbook of Inductively Coupled Plasma Mass Spectrometry*, Blackie, Glasgow and London, UK. ISBN 0–216–92912–1.

Jeannot, R., Lemière, B. & Chiron, S. (2001) doc BRGM, 298, Eds BRGM, Orléans.

Kaerstensen, K.H. (1996) Nordic guidelines for chemical analysis of contaminated soil samples, NT Techn Report 329.

Kersten, M. & Förstner, U. (1995) In *Chemical Speciation in the Environment*, Ure, A.M. and Davidson, C.M. (eds), Blackie Academic Professional, Glasgow, 234–275. ISBN 0–87371–697–3.

Kibblewhite, M. (2001) Identifying and dealing with contaminated land, In *Assessment and Reclamation of Contaminated Land*, Hester and Harrison (eds), **16**, Royal Society of Chemistry (RSC). ISBN 0–85404–275–X.

Kingston, H.M. & Walter, P.J. (1992) Comparison of microwave versus conventional dissolution for environmental applications, *Spectroscopy*, **7**(9), 20–27.

Kingston, H.M. & Haswell, S. (eds) (1997) Microwave Enhanced Chemistry: Fundamentals, Sample Preparation and Applications, *America Chemical Soc.*, Washington, D.C.

Lamble, J.K. & Hill, S.J. (1998) *Analyst*, **123**, 103R–133R.

Majors, R.E. (1991) An Overview of Sample Preparation, LC-GC, 9/1.

Quevauviller, P., Mayer, E.A. & Griepink, B. (1995) *Quality Assurance for Environmental Analysis*, Quevauviller/Mayer/Griepink (eds), **17**, 63–88. Elsevier Science, ISBN 0–444–89955–3.

Quevauviller, P. (1998) *Method Performance Study for Speciation Analysis*, The Royal Society of Chemistry (RSC). ISBN 0–85404–467–1.

Séby, F., Potin-Gautier, M., Lespés, G. & Astruc, M. (1997) *The Science of Total Environment*, **207**, 81–90.

Slavin, W., Carnrick, G.R. & Manning, D.C. (1984) *Anal. Chem.*, **56**, 163–168.

Standard Methods for Examination of Water and Wastewater (1998) 20th Edn, APHA AWWA WEF, 3-1 3-105. ISBN 0–87553–235–7.

SW-846 EPA Method (2000) Test methods for evaluating solid waste, Rev. 5, Environmental Protection Agency, Washington, DC.

SW-846 EPA Method 3050B (1998) Acid digestion of sediments, sludges, and soils, in Test Methods for Evaluating Solid Waste, Rev. 2, Environmental Protection Agency, Washington, DC.

SW-846 EPA Method 3051A (1998) Microwave assisted acid digestion of sediments, sludges, and soils, in Test Methods for Evaluating Solid Waste, Rev. 0, Environmental Protection Agency, Washington, DC.

SW-846 EPA Method 3052 (1998) Microwave assisted acid digestion of siliceous, and organically based matrices, in Test Methods for Evaluating Solid Waste, Rev. 0, Environmental Protection Agency, Washington, DC.

SW-846 EPA Method 3060A (1998) Alkaline digestion for hexavalent chromium, in Test Methods for Evaluating Solid Waste, Rev. 1, Environmental Protection Agency, Washington, DC.

SW-846 EPA Method 7000A (1992) Atomic absorption methods, in Test Methods for Evaluating Solid Waste, Rev. 1, Environmental Protection Agency, Washington, DC.

Swedish Standard Institution SSI (1997) Soil analysis – determination of trace elements in soils – extraction with nitric acid.

Templeton, M.D., Ariese, F., Cornelis, R., Danielsson, L.G., Muntau, H., van Leeuwen, H.P. & Lobinski, R. (2000) *Pure Applied Chemistry*, **72**, 1453–1470.

Thomas, P. (1996) *Spectra Analyse*, **189**, 27.

Thomas, P., Finnie, J.K. & Williams, J. (1997) *J. Anal. At. Spectrom.*, **10**, 615.

Thomas, P. & Morin, J.M. (2000) *Spectra Analyse*, **216**, 25.

Thompson, M. & Walsh, J.N. (1983) *Handbook of Inductively Coupled Plasma Spectrometry*, Blackie, Glasgow and London, UK. ISBN 0–216–91436–1.

Todoli, J.L., Gras, L., Hernandis, V. & Mora, J. (2002) *J. Anal. At. Spectrom.*, **17**, 142.

Welz, B. & Sperling, M. (1999) *Atomic Absorption Spectroscopy*, Ind. edn. VCH, Federal Republic of Germany. ISBN 3–527–28571–7.

5   Analysis of inorganic parameters

George E. Rayment, Ross Sadler, Andrew Craig,
Barry Noller and Barry Chiswell

5.1   Introduction

The parameters covered in this chapter are diverse. Although described here as inorganic, pH and redox potential of contaminated land are likely to be as much a product of organic processes as inorganic ones. Furthermore, while much of the interest in how parameters except asbestos transform and are analysed is associated with their behaviour in aqueous solution, the influence of organic species upon any changes cannot be ignored. Without an appreciation of such likely transformations, the analyst will often be unable to perform the appropriate analysis and to prevent serious mistakes associated with interferences. Inorganic analyses of parameters such as pH, electrical conductivity and redox potential are among the most basic and important analyses required when assessing any potentially contaminated land. The physical information obtained from such measurements will to some extent define the boundaries of the speciation analysis to follow. It is therefore of prime importance to determine these parameters correctly. Many of the methods are techniques of partial extraction and so are empirical. The results obtained will be dependent on the method employed, and the method should therefore be closely followed. Small deviations can produce a significant difference in the results.

Choice of method should take account of likely knowledge gain, accuracy and precision, ease of use, time and resources necessary to perform the test/s, cost per sample, operator safety, and the nature and disposability of environmental residues generated during the testing phase, including releases to air and water. This chapter reviews a range of inorganic methods of analysis applicable to many applications. Each covers the extraction (or digestion) of the inorganic constituent from its matrix, through to the determination once extracted, taking account of interfering substances when these are likely to be significant. Table 5.1 summarises many of these methods.

5.2   Electrical conductivity

The electrical conductivity (EC) of a soil suspension provides an estimate of the concentration of soluble salts in the soil; in most soils these consist predominantly of the cations Na^+, Mg^{2+} and Ca^{2+} and the anions Cl^-, SO_4^{2-}

Table 5.1 Summary details of test methods in this chapter, including typical concentration ranges

Analyte	Usual sample size (g)[a]	Extractant	Sample/solution ratio	Typical extraction time	Analytical finish	Typical concentration range	Expected interlab precision
Electrical conductivity (EC)	10–20	Water	1:25 or 1:5	60 minutes	Conductivity cell	0.001–5 dS/m	CV = 6%
pH	10–20	Water or salt solution	Define e.g. 1:5	60 minutes	Electrode	1–13	CV = 4%
Redox potential	*In situ*	n.a.	n.a.	n.a.	Redox electrode	n.a.	CV = 10%
Soluble chloride (Cl)	As for pH	Water	1:5	60 minutes	• Electrochemical • Colorimetric • IC	1–2000 mg/kg	CV ≤ 10%
Soluble boron (B)	10	Hot water	1:2	10 minutes	• Colorimetric • ICPAES	0.1–50 mg B/kg	CV = 40–50%
Total sulfur	n.a.[b,c]	n.a.	n.a.	n.a	XRF	0.01–5% S	CV = 10–15%
Soluble sulfate (SO$_4$-S)	20	Water	1:5	17 hours	• IC • ICPAES (soluble S)	1–50 mg S/kg	CV = 15–30%
Chromium reducible S	1.0[d]	CrCl$_2$	n.a.	20 minutes	Distillation/titration	0.01–5% S	CV = 10–25%
Soluble cyanides (CN, CNO, SCN)	10–20	Water	1:5	60 minutes	• Colorimetric • IC	1–1000 mg/kg	CV = 10–25%
Total cyanide (CN)	10–20	Acid	n.a.	60 minutes	• Colorimetric • IC	1–1000 mg/kg	CV = 10–25%

[a] particle size of <2 mm air dry (40°C) unless otherwise stated.
[b] (n.a.) not applicable.
[c] particle size <2 μm dried at 65°C.
[d] particle size <0.5 mm dried at 85°C.

and HCO_3^-. Typical soil:water ratios (deionised or distilled) are 1:1, 1:2, 1:2.5 and 1:5. The 1:5 ratio is preferred as it gives an approximation of soil ionic strength (Gillman & Bell 1978). The same extract can also be used to determine pH and water-soluble chloride and nitrate. Even that ratio is insufficient to solubilise more than about 1% of gypsum, which yields an EC of about 2 dS/m. Electrical conductivity values increase with increasing temperature and must be corrected if not measured at 25°C (APHA 1998).

5.2.1 Soil:water extract

For a 1:5 soil:water suspension, weigh 20.0 g air-dry soil (<2 mm) into a suitable bottle or jar and add 100 ml deionised water. Mechanically shake (end-over-end preferred), at 25°C in a closed system for one hour to dissolve soluble salts. Subsequently, allow around 20–30 minutes for the soil to settle. Subsequent actions depend on which measurements are to be made.

5.2.2 EC measurement and reporting

Initially calibrate the conductivity cell and meter in accordance with manufacturer's instructions. When calibrated, dip the conductivity cell into the supernatant, moving it up and down slightly without disturbing the settled soil. Take the reading with the cell stationary, when the system has stabilised. Rinse the cell with deionised water between samples and remove excess water. Complete EC measurements within three to four hours of obtaining the aqueous supernatant. Report EC (dS/m) at 25°C on an air-dry basis. No correction for moisture present in the sample is necessary.

5.3 pH

Soil pH reflects the intensity of acidity that in turn influences soil conditions and plant uptake of metal(loid) contaminants. pH influences solubility and activity of various biologically important elements and processes. The result is affected by the soil/liquid ratio and by the composition and temperature of the equilibrating solution. Typically, soil pH values increase as the ratio of soil/liquid widens. Salt solutions, in addition to water (pH_w), such as 0.01 M $CaCl_2$ and 1.0 M KCl are often used to *buffer* for differences in hydrogen ion activity caused by changes in soil ionic strength. Thus, these extractions are used in soils that have varying levels of soluble ions (e.g. following fertiliser application, rainfall, etc.). Differences are commonly 0.5–1.0 pH unit lower than for a 1:5 soil:water extract, except when the soils are highly weathered (Conyers & Davey 1988; Ahern *et al.* 1995). For highly weathered soils, the differences between water and salt-solution pH tests can be insignificant or even reversed. Values can vary due to the choice and position of the electrode array and whether or not measurements are made in the supernatant or in a stirred soil/liquid suspension.

The pH of surface soils (0–100 mm) commonly range from 6.0 to 8.0, and it is useful to note that pH_w values around 4.0 or less suggest the presence of sulfides, while levels in excess of 8.5 are indicative of the presence of significant quantities of exchangeable Na^+.

Measurement of pH involves detection of the potential of a glass-calomel electrode array (separate electrodes or in combination) using a pH or millivolt meter, standardised against buffer solutions of known pH. Values for these buffers vary with temperature (Bates 1962; Alvarez 1984).

Procedures based on *extraction* with deionised water, dilute calcium chloride (0.01 M $CaCl_2$) potassium chloride (1 M KCl), saturated sodium fluoride (NaF) and hydrogen peroxide (H_2O_2) have been described. All procedures (except for H_2O_2, which ideally uses field-moist soil) are based on the use of air-dry soil of <2 mm particle size. No correction for water content is made when reporting pH results. The soil/solution ratio and the temperature of measurement should be noted on the report, as other soil/solution ratios are often used. These include saturation extracts and soil/solution ratios of 1:1 and 1:2.5.

5.3.1   pH of 1:5 soil/water suspension (pH_w)

A 1:5 soil/water extract is mechanically stirred during measurement to minimise changes in electrode potential associated with suspension effects and positioning of electrodes. Results by this procedure are commonly higher by about 0.5–0.6 of a pH unit (Baker *et al.* 1983) than those measured in the field by the mixed indicator/barium sulfate method of Raupach and Tucker (1959). The water should have a pH ≥ 6.5 but ≤ 7.5. If necessary, boil distilled or deionised water for 15 minutes and cool under CO_2-free conditions. EC should be $<10^{-3}$ dS/m. Standardisation of the equipment is undertaken using standard pH solutions; usually pH 4.00, 7.0 and 9.183. Such solutions are normally purchased ready for use.

5.3.2   pH of 1:5 soil/0.01 M calcium chloride extract (pH_c)

This method uses dilute calcium chloride (0.01 M $CaCl_2$) at a 1:5 soil/solution ratio. Because of trends towards robotics, the soil/0.01 M $CaCl_2$ extract is stirred during measurement. A typical testing range is ≈ 1.5–11.

5.3.3   pH of 1:5 soil/1 M potassium chloride extract (pH_K)

A potassium chloride solution (1 M KCl) is used in this method, with pH determined under conditions similar to those for 0.01 M $CaCl_2$. This method, which is particularly applicable to acidic, highly weathered soils dominated by colloids with variable surface charge characteristics, may be used for comparison with results obtained in Method 5.3.1.

5.3.4   pH of sodium fluoride suspension (pH$_F$)

This soil pH$_F$ procedure is used to assess the presence of active aluminium. When immersed in a solution of NaF, active Al adsorbs F$^-$ ions with a consequent release of OH$^-$, leading to higher pH values. High pH values in NaF are found in soils derived from volcanic ash and in illuvial horizons of podzolised soils (Fielders & Perrott 1966).

5.3.5   pH of hydrogen peroxide extract (pH$_{HO}$)

This method, adapted from Ford and Calvert (1970) and Rayment and Higginson (1992) is mostly used on *as received* (field moist) soil to obtain an indication of the presence of oxidisable sulfide ions (S^{2-}). It can also be used on soils dried rapidly at 80–85°C to *stabilise* pyritic soils for later analysis.

If present, S^{2-} ions are oxidised with heated 30% H$_2$O$_2$ to form H$_2$SO$_4$, with a consequential lowering of pH. Should soil pH decline to 3.5 or less, it can usually be assumed that soil and drainage water could become very strongly acidic if subsequently exposed to the atmosphere following draining or land levelling. A measured or visible increase in the sulfate content of the unreacted (water only) and reacted extract provides useful confirmation that the pH shift is due the presence of S^{2-} ions rather than organic matter.

5.3.5.1   Procedure

Weigh 1 g soil (usually field moist) into a 200 ml, wide-mouth conical flask. Add 20 ml of 30% analytical grade H$_2$O$_2$ and place on a boiling water bath. Remove from the water bath if the reaction is too vigorous. Otherwise, continue heating until the solution clears, indicating that the reaction is complete. This should take at least 30 minutes, and all organic matter should have reacted. Should the solution fail to clear on completion of the initial reaction, add additional 30% H$_2$O$_2$ (with care) and continue heating until clearing is achieved. Allow to cool to room temperature, standardise the pH meter and measure pH as for Method 5.3.1. The presence of SO$_4^{2-}$ in the oxidised solution can be confirmed qualitatively by the addition of 1 ml of 10% BaCl$_2$ solution to the filtered or centrifuged supernatant. If present, a visible precipitate of BaSO$_4$ will form. Report pH$_{HO}$, noting the moisture status of the sample.

5.3.6   ΔpH

This calculation provides some indication of the sign of the electrical charge that currently dominates in the soil (Rayment & Higginson 1992). This calculation is most applicable to highly weathered acidic soils with variable-charge characteristics. A typical reporting range is ≈ −1.5 through to +1.5, with negative values being the most common.

Calculation

$$\Delta pH = pH_k - pH_w$$

where pH_K refers to Method 5.3 and pH_w refers to results obtained by Method 5.3.1. Report ΔpH, including the $-$ve or $+$ve sign.

5.4   Redox potential

5.4.1   Summary

Redox potential (Eh) characterises the oxidation–reduction condition of a soil. This in turn provides a means to assess soil genesis, soil fertility and status of soil contaminants (Liu & Yu 1984). Redox potential is well known to be difficult to measure precisely with conventional methods, for reasons ranging from slow electrode response to soil condition, especially for poorly poised (redox capacity) soils (Bohn 1971; Ponnamperuma 1972). Usually measurements are made under specified conditions, with values being dependent on the experimental conditions. As the measured value is conditional, it may not be sufficiently precise for some physicochemical studies. Nevertheless, there is unique information about soil condition that can be derived by measuring the redox potential.

5.4.2   Capacity factor (Poise)

Redox reactions describe the coupling of reduction reactions to oxidation reactions in free soil. Electrons are transferred from reductors (e^- donors or reducing agents) to oxidants (e^- acceptors or oxidising agents) (James & Bartlett 2000).

   Generally for a reduction half-reaction: $ox + ne^- = red$, the electron activity, $(e^-) = [(red)/K(ox)]^{1/n}$.

   Electron activity is a function of the ratio of reduced-to-oxidised species activity and of the equilibrium constant (or Gibbs free energy) for the reaction, since: $\Delta G_T^0 = -RT \ln K$.

   Two concepts having similarity of response from soils are those of *proton activity* and *electron activity* as the thermodynamic activity is defined as the ratio of H^+ and e^- activity under standard state conditions (James & Bartlett 2000).

   The capacity factor in redox is referred to as *poise* and is defined as the change in added equivalents of reductant or oxidant to bring about a one unit change in pe (or Eh change of 59 mv). The concept is similar to that of buffer capacity for pH (Stumm & Morgan 1996). However, poise in soils has been less studied than pH buffering.

5.4.3   Usefulness of pe

The measurement of pe together with pH can describe the equilibrium condition of a soil through the master variable, pH and pe and their sum (Lindsay

1979). Individual pe–pH relationships give a straight line in which log K and the activities of the oxidant and reductant determine the y intercept; the ratio of H^+ to e^- consumed determines the slope (James & Bartlett 2000). The effects of pH on each redox couple can be clearly shown, as the interactions of master variables are delineated. The usefulness of the interactions of pe and pH can be extended by incorporating algorithms for redox reactions to estimate pe, pH and reductant and oxidant activities in conjunction with dissolution and exchange equilibria using multi-equilibria programs such as GEOCHEM (Allison & Brown 1995). This kind of program utilises the minimisation of the Gibbs free energy of a system of multiple chemical equilibria to predict the preferred reaction products. The ordering process of relative free energies identifies the free energy chemical formations, which take into account pe, pH and reductant and oxidant activities in conjunction with dissolution and exchange equilibria. GEOCHEM, in particular, incorporates equations for redox reactions (Allison & Brown 1995).

5.4.4   Measurement of soil redox and electrodes

The most common method for quantifying electron activity of soils is to measure the potential difference between a Pt indicator electrode and a calomel or Ag/AgCl reference electrode, both connected to a voltmeter (James & Bartlett 2000). This method assumes that the Pt electrode is inert and does not react chemically while achieving equilibrium with the electroactive species in the soil and on soil colloids. During the measurement of redox potential, it is important to minimise electrode polarisation caused by current flowing through the electrode (Aomine 1962). Generally, the potentiometric measurement is unreliable for accurate assessment of redox status, especially of aerobic soils. Methods employing analysis of soil solution analytes, indicative of redox status, may be more reliable. A new procedure is proposed by James and Bartlett (2000) to evaluate electron activity for soils by using a combination of analytical and thermodynamic approaches. For example, manganese and iron concentrations and their associated mineralogies are easily measured by a variety of methodologies, giving a basis for assessing pe for manganese and iron oxide-dominated systems (James & Bartlett 2000). In contrast, dissolved gases are harder to measure accurately and qualitative estimates may be sufficient to obtain accurate evaluation of pe. The key step to accurate estimation of pe is therefore knowledge of the contribution of redox couples to electron activity.

Platinum and suitable reference electrodes are relatively easy and inexpensive to construct, but the measurement technique may significantly alter measured voltages. Quantification of H_2 may be more reliable than Eh measurements for identifying anoxic redox process. Quantification of H_2 may be achieved by using a gas-stripping procedure by N_2 displacement following sample collection (Chapelle *et al.* 1996). However, because of complexities in interpretation, using H_2 measurements as a sole indicator of redox processes is not appropriate. A more reliable procedure is to interpret H_2 concentrations in the context of electron-acceptor availability.

The limitations have been described (James & Bartlett 2000), especially with respect to how long the electrodes can be left in place in the soil and with respect to interpreting the Eh values obtained. A specific limitation relates to the inadequacy of Pt electrode potential measurement.

Inadequacies of the Pt electrode potential may arise as follows:

1. Unreliable dissolved oxygen status.
2. Irreversibility of redox couples.
3. Mixed potentials.
4. Coupling of pH and pe.

Other electrodes which have been examined are as follows: graphite electrodes, simple Ag, AgCl reference electrodes (Farrell *et al.* 1991); microelectrodes constructed from micropipets (Pang & Zhang 1998; Shaikh *et al.* 1985). James and Bartell (2000) also list several general reviews and examples of redox measurements.

5.5   Water soluble chloride

Chloride ions are released during the natural attenuation and interventionist degradation of chlorinated organic compounds. Being soluble and persistent it can be used to identify the extent of groundwater contamination plumes. Chloride (Cl^-) is a common monovalent ion in soils, particularly in semi-arid areas of former marine influence. It also tends to build up in many irrigated soils due to rises in groundwater and as a consequence of evaporation. It is essential for plant growth but rarely deficient. Chloride readily forms complexes with metals such as cadmium (Cd), thus enhancing the solubility and plant availability of Cd across a wide pH range. Chloride is also important in land-use assessment due to its possible accumulation in soil profiles and waters to levels detrimental to plant growth. Typical concentrations in soil range from almost zero to well in excess of 1000 mg/kg, while soil solution concentrations vary from around zero to 32 mM. Symptoms of Cl toxicity in plants include burning of leaf tips or margins, bronzing and premature yellowing of leaves, and leaf drop. Values of water-soluble Cl useful for the general interpretation of soil chemical analyses are: <100, 100–300, 300–600, 600–2000 and >2000 mg/kg for ratings of very low, low, medium, high and very high, respectively (Rayment & Bruce 1984).

Analytical estimation of Cl in soil suspensions can be performed (following filtration or centrifugation) by potentiometric titration, automatic colorimetric analysis (automated ferricyanide method) and ion chromatography. Choice of method is largely an issue of laboratory preference and the level of instrumental sophistication available.

5.5.1 Chloride – 1:5 soil/water extract, ion chromatography (chemical suppression of eluent conductivity)

Ion chromatography permits quantitative determination of common anions (F, Cl, NO_2, Br, NO_3, HPO_4, SO_4) in known aqueous matrices within the mg/l concentration range. The technology can effectively distinguish the halides (F, Cl, Br) and eliminates the need to use hazardous reagents during analysis (APHA 1998). Ion chromatography analyses can be completed sequentially within 30 minutes or less using a few ml of extract. If a particular anion is present at a very high concentration, it may interfere by causing a very large peak on the chromatogram, causing other peaks to be masked. Such interferences are often avoided by dilution and/or via the use of specialised columns.

Determination of Cl concentration – in conjunction with the other anions mentioned – involves injecting a finely filtered aliquot (<0.45 μm with <0.2 μm membrane filter preferred) of the 1:5 soil/water extract into an ion chromatograph appropriately configured. The method described is based on O'Dell *et al.* (1984) and Rayment and Higginson (1992). Refer to Australian Standard AS 3741 (ANON 1990), Section 4110 of APHA (1998) and Method D 4327-97 of ASTM (2000) for further details on analysis by ion chromatography.

5.6 Water soluble boron

Hot water extraction of soil boron (B) (Berger & Truog 1939) has been used widely to obtain an index of soil B status. The method presents difficulties for routine analysis as it is often difficult to prepare water extracts free of colloidal material. Dilute salt solutions have largely overcome this limitation, two examples being the use of 0.01 M $CaCl_2$–0.05 M mannitol (Cartwright *et al.* 1983) and hot 0.01 M $CaCl_2$ (Aitken *et al.* 1987). The empirical method described is based on the use of hot water reflux for soil extraction, with alternative analytical finishes based on a colour reaction and ICPAES.

5.6.1 Hot water extractable soil B

The hot water extractable soil B method is based on refluxing air dry soil for 10 minutes with deionised water at a 1:2 soil/solution ratio. After replacing liquid lost during reflux, extracts are filtered while hot and allowed to cool. The first of the two analytical finishes is a colorimetric procedure which utilises azomethine-H (Rayment & Higginson 1992). It is preferred when ICPAES is not available. Other colorimetric procedures for determining B, including carminic acid (Hatcher & Wilcox 1950) and curcumin (Hayes &

Metcalfe 1962) need concentrated acids or organic solvents. The likelihood is that ICPAES will yield higher results than azomethine-H.

5.7 Sulfur

Total sulfur (S) concentrations in soils vary widely from close to undetectable levels to >10 000 mg/kg on a dry basis, with a mean of around 430 mg/kg (Rayment & Higginson 1992). Many organic and inorganic fractions of sulfur exist (Williams 1975). In general, organic S fractions are highly correlated with organic carbon (C) and total nitrogen (N), with the C:N:S ratios of natural soils of humid regions commonly around 140:10:1.3 (Probert & Samosir 1983).

Solid phase water-soluble sulfates, especially gypsum ($CaSO_4 \cdot 2H_2O$), are likely to occur in low rainfall areas with a prior marine influence. For example, it is found in many soils of inland Australia (Rayment et al. 1983).

Actual and potential acid sulfate soils (ASS) are another class of soils that contain elevated concentrations of total S as sulfide. These require careful management as on oxidation, strong sulfuric acid (H_2SO_4) can form, releasing acid, iron, aluminium and sometimes heavy metals to nearby waterways, which is environmentally unacceptable. The pyrite (FeS_2) in these soils is typically associated with Holocene marine sediments deposited during the last 6000–12 000 years, when there was flooding of coastal embayments and estuaries and the drowning of river valleys. The sulfide is often in excess of the available Fe and this excess usually produces the unpleasant sulfurous odour of H_2S when the sediments are disturbed. Elemental sulfur is most commonly found in volcanic soils.

Measurements of total S are undertaken to assess the size of the soil S pool. In general, total S is poorly correlated with plant response to applied S fertiliser, except when soil S reserves are very low. Total S is also a poor indicator of a soil's tendency to corrode concrete, but is a useful early screening tool for the possible presence of pyrite and other sulfides in ASS. An X-ray fluorescence method for total soil S is described. It is based on use of dry-powder pellets.

Chromium reduction to measure reduced inorganic sulfur compounds in sediments was proposed by Zhabina and Volkov (1978). Since then it has found wide use internationally (Sullivan et al. 1999), particularly when pyritic sediments and acid volatile mono-sulfides are expected. The method is not measurably affected by sulfur in organic matter or sulfates (Canfield et al. 1986; Morse & Cornwell 1987). Accordingly, this chromium reducible sulfur method (S_{Cr}) is especially useful on samples with appreciable organic matter and also for sandy soils where the %S action criterion is very low (e.g. as low as 0.03%S).

Sulfate is the most common water-soluble form of S in soil. It does not follow, however, that all sulfate can be extracted with water, even when the solubility product is not exceeded (see Barrow (1975) for information on factors affecting adsorption of SO_4^{2-} by soils, as well as its measurement). Particularly in acidic soils dominated by colloids with constant surface potential, sulfate may be held on anion exchange sites and on other surfaces with an affinity for sulfate ions. For this reason, a variety of aqueous, dilute salt solutions based on chloride, nitrate and phosphate ions have been used to enhance the extraction of sulfate. Here it is assumed that the soils in question contain enough sulfate-S to be sufficiently but not quantitatively extracted with water. Accordingly, variables such as temperature and the soil to solution ratio influence the result obtained.

5.7.1 Total sulfur – X-ray fluorescence

This X-ray fluorescence (XRF) method is based on techniques described by Brown and Kanaris-Sotiriov (1969), Norrish and Hutton (1964) and Norrish and Chappell (1967). Dry-powder pellets of finely divided (<2 μm) soil are used.

5.7.1.1 Preparation of standard pellets

Prepare solid standards of known S content by adding gypsum (AR $CaSO_4 \cdot 2H_2O$), and appropriate $(NH_4)_2SO_4$ solutions of known concentration to weighed quantities of silica (purchased, or prepared as described by Hewitt (1965)). These standard pellets have an extended life.

Note: Distribute the gypsum or S standard evenly or dropwise over the surface of the silica without wetting any part of the beaker. Dry the impregnated silica (65°C), grind to <2 μm and press into a pellet using a hydraulic press of around 25 tonne total force. Typically pellet about 2 g *standard* plus H_3BO_3 mix into a 45 mm diameter disc with a H_3BO_3 backing.

5.7.1.2 Preparation of soil pellets

Oven dry (65°C) approximately 10 g air-dry soil (<2 mm), add 0.5 g technical grade H_3BO_3 to serve as a binder, place into a clean 100 g capacity ring and pluck head, and grind in a *shatterbox* for a minimum of 2 minutes, or until soil particle size is <2 μm. Next pellet about 2 g soil plus H_3BO_3 mix (<2 μm) into a 45 mm diameter disc with a H_3BO_3 backing, using a hydraulic press of around 25 tonne total force.

5.7.1.3 Analysis

Follow manufacturer's recommendations for matrix correction(s) and measurements. When feasible make the nett count (i.e. peak-background) equal to the concentration of the standard pellet (or related to it by some power of 10), so that element concentration may be recorded directly from the display. Report total S (%) on an oven-dry basis.

5.7.2 Pyrite and other iron disulfides and acid volatile sulfides (chromium reducible sulfur)

This method by Sullivan *et al.* (2000) is based on the conversion of reduced inorganic sulfur to H_2S by hot acidic chromium(II) chloride ($CrCl_2$) solution. The hydrogen sulfide (H_2S) evolved is trapped in a zinc acetate solution as ZnS, and subsequently quantified by iodometric titration (see Fig. 5.1). The reduced inorganic sulfur compounds measured are: (i) pyrite and other iron disulfides; (ii) elemental sulfur; and (iii) acid volatile sulfides (e.g. greigite and mackinawite). The method can be made specific to the iron disulfide fractions if pre-treatments are used to remove the acid volatile sulfides and elemental S fractions.

The main difference in this method compared to that of Sullivan *et al.* (1999) is in the shorter reaction time of 20 minutes compared to the original reaction time of one hour. Suitable sulfide standards should be used frequently to determine the recovery rates, with analyses corrected accordingly. Sullivan *et al.* (1999), using a 5.0% pyrite standard had a pyritic recovery of 98.7%. Canfield *et al.* (1986) recorded a recovery of 95.9%, whereas Raiswell *et al.* (1988) reported a recovery rate 'close to 100%'.

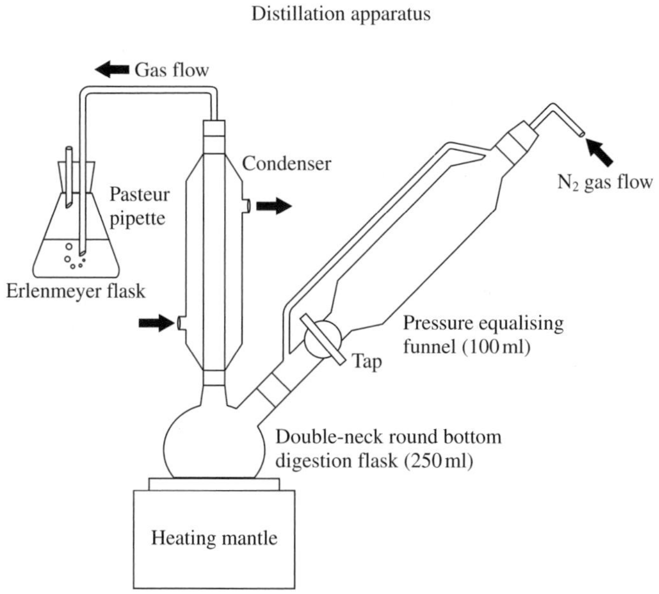

Distillation apparatus

Figure 5.1 Schematic of the apparatus used in the chromium reducible method for determination of S_{Cr}. Nitrogen (N_2) gas flow rate is regulated at 2–3 bubbles/second. Allow the N_2 gas to purge the system for about 3 minutes between samples.

5.7.2.1   Procedure

Weigh $1.0\,g^1$ of dried[2] (85°C) soil of <0.5 mm into a double-neck round-bottom digestion flask. Add 2.059 g of chromium powder and then 10 ml ethanol (95% concentration) to the digestion flask and swirl to wet the sample. Place the digestion flask in the heating mantle and connect to the condenser, with all apparatus in an efficient fume cupboard.

Attach the pressure equalising funnel making sure the gas-flow arm is facing the condensers and the solution tap is shut. Attach a Pasteur pipette to the outlet tube at the top of the condenser and insert it into a 100 ml Erlenmeyer flask containing 40 ml of zinc acetate solution (dissolve 60 g of zinc acetate in 1.5 l of deionised water; add 200 ml of 28% analytical grade ammonia solution and make to 2 l with deionised water).

Turn on the water flow around condenser and make sure that all ground glass fittings are tight (any lubrication should be S-free). Add 60 ml of 5.65 M HCl to the glass dispenser in the pressure equalising funnel. Connect the N_2 flow to the pressure equalising funnel and adjust the flow to obtain a bubble rate in the zinc acetate solution of about 3 bubbles/second. Allow the N_2 gas to purge the system for about 3 minutes, then slowly release the 5.65 M HCl from the dispenser. (*Note*: The 5.65 M HCl should be added to the sediment chromium power very slowly.) This procedure must be carried out in a fume cupboard.

Wait for 2 minutes before turning on the heating mantle and adjust the heat so that a gentle boil is achieved. Check for efficient reflux in the condenser. Allow to digest for 20 minutes, then remove the Erlenmeyer flask and wash off any ZnS on the Pasteur pipette into the Erlenmeyer flask with water from a wash bottle.

Add 20 ml of 5.65 M HCl and 1.0 ml of starch indicator solution to the zinc acetate solution and gently mix on a magnetic stirrer. Titrate the zinc acetate trapping solution with standard iodine solution (22.5 g of KI and 3.2 g I_2 in 1 l deionised water) to a permanent blue end-point.

5.7.2.2   Calculation of the chromium reducible sulfur content

Oxidation of iodine (to tetrathionate) involving one mole of thiosulfate corresponds to one *equivalent* of iodine (i.e. ½ I_2). The relevant chemical reaction in the absence of sodium carbonate or equivalent is:

$$S^{2-} + I_2 \rightarrow S + 2\,I^-$$

$$2S_2O_3^{2-} + I_2 \rightarrow S_4O_6^{2-} + 2\,I^-$$

This is given below

$$\%S_{Cr}(\text{oven dry}) = \frac{(A-B) \times C \times 1600}{\text{Mass of soil (mg)}} \times 2$$

where:

A = The volume of iodine (in ml) used to titrate the zinc acetate trapping solution following the soil digestion,

B = The volume of iodine (in ml) used to titrate the zinc acetate trapping solution following a blank digestion,

C = The molarity of the iodine solution as determined by titration with standard 0.025 M $Na_2S_2O_3$ solution, noting that $C = [(0.025 \times D)/E]$, where D = titre (ml) of standard 0.025 M sodium thiosulfate and E = volume (ml) of iodine solution titrated.

Report % S_{Cr} on an oven dry (85°C) basis.

5.7.3   Water extractable sulfur – ICPAES

Sulfate S is extracted from air-dry soil of <2 mm particle size with deionised water, using a soil to solution ratio of 1:5 and an extraction time of 17 hour at 25°C. This extracting solution will not displace adsorbed S, and will not necessarily dissolve all the gypsum that could be present. The extracted S is then determined in an aliquot of clear soil extract by inductively coupled plasma atomic emission spectrometry (ICPAES). In conjunction with vacuum optics, ICPAES is an efficient technique for the measurement of S in soil extracts. At the wavelength, 182.036 nm, there is virtually no interference from Ca^{2+}.

5.7.4   Water extractable sulfate-sulfur – ion chromatography (chemical suppression of eluent conductivity)

The equipment and issues pertaining to this analysis are similar to those for soil chloride (see Method 5.5.1) using ion chromatography. The method is specific to sulfate, whereas ICPAES measures total S in the aqueous extracts. Accordingly, results will usually be lower for this analytical finish compared with Method 5.7.3. The only post-extraction preparation necessary, other than dilution, is the removal by filtration of particulate matter >0.20 µm. Since the concentration of SO_4^{2-}-S can be influenced by biological transformations, aqueous soil extracts should be kept at 4°C if there is any delay (e.g. overnight) between completion of extraction and analysis. Under such conditions filtered aqueous samples should remain stable for at least 28 days (O'Dell et al. 1984). The analysis can be undertaken in sequence with other anions (Dick & Tabatabai 1979; O'Dell et al. 1984).

5.8   Cyanides, thiocyanates and cyanates

5.8.1   Cyanides

Technically, a cyanide is any chemical that contains the C≡N group. Thus, the family includes chemicals such as hydrogen cyanide, simple inorganic cyanides, complex metal cyanides, acrylonitrile and cyanogenic glycosides. Although a few of these substances (e.g. cyanogenic glycosides) occur in nature, the majority result from anthropogenic activities. In terms of soil contamination, anthropogenic activities are exclusively responsible for the presence of cyanide in polluted soils. The major sources of cyanide at contaminated sites are as follows:

1. Wastes from gold mining operations in which cyanidation has been used to extract gold from the ore. Mine tailings from these operations contain a significant quantity of free or available cyanide, and other complexes including cyanide degradation compounds. Cyanidation processes work on the basis of formation of the strong complex $Au(CN)_2^{2-}$, which has a formation constant at least two orders of magnitude higher than most other metals; the gold cyanide complex is selectively adsorbed by activated carbon. This, however, does not mean that other metal cyanide complexes are not formed in the process. Elements associated with gold in sulfide ores include iron, cobalt, nickel, gold, silver, mercury, copper, lead, zinc, antimony and bismuth, but particularly copper and iron, which are most common. Thus the potential exists for the formation of a mixture of metallocyanide complexes, which occur at environmentally significant concentrations in the waste. At least some of these complexes are readily broken down into free cyanide and their presence in the waste is thus of potential concern (Smith & Mudder 1991). Iron and cobalt complexes can be broken down photolytically which maybe controlled by the presence of sunlight (Johnson *et al.* 2002). Copper cyanide complexes present a toxic hazard to wildlife (Donato 1999).

2. Wastes from purifiers at manufactured gas plant (MGP) sites. Cyanide is a by-product of coal retorting and is/was trapped along with sulfide in the iron oxide purifiers at coal gas plants. The total content of cyanide in MGP wastes is generally 1–2% (although the levels may go much higher in some cases), of which more than 97% is in the form of complex iron hexacyanides (Köster 2001).

3. Blast furnace sludge resulting from production of pig iron. In this process, sodium and potassium cyanides are generated in the blast furnace and leave as dust in the top gas. This top gas is subjected to wet scrubbing, which results in a muddy cyanide containing waste (Mansfeldt & Biernath 2001).

4. Wastes from metallurgical processes. A number of metallurgical processes, notably case hardening and electroplating, make use of alkaline baths containing simple inorganic cyanides. In larger operations and developed countries the cyanide is generally recycled. Historically, this was not always the case and the situation persists in developing countries to this day. Thus, the waste disposal areas, which were usually proximal to the operation, are contaminated with cyanide. Depending on the content

of metal ions (e.g. iron) in the soil that would complex the cyanide into an inert form, the contamination may present a risk to public and environmental health.

5. Wastes from the photographic industry. The photographic industry makes use of a range of cyanides, from ferrihexacyanides to aqueous sodium cyanide. These may also pose an environmental threat in the case of disposal, although recycling is generally a more common fate for these chemicals (Köster 2001).

In addition to these, there exists potential for cyanide to enter the soil through other, as yet unevaluated pathways. For example, the processing of cyanide-containing plant material results in waste streams containing relatively high levels of cyanide and this may be a problem, particularly in tropical countries (Siller & Winter 1998).

As regards determination of cyanides, environmental studies are almost entirely concerned with the acute toxicity of cyanide. Although cyanide can exist in many chemical forms, hydrogen cyanide is the most toxic. The reaction:

$$HCN \rightleftarrows H^+ + CN^-$$

has a pK_a of 9.36. Hence determinations of cyanide in environmental media relate either directly to hydrogen cyanide or to cyanide compounds that liberate hydrogen cyanide under the mild acidic conditions of the stomach. This includes both simple inorganic cyanides and weak cyanide complexes. The accompanying table illustrates the relative toxicity of various inorganic cyanide species. Although not generally encountered in environmental media, cyanogenic glycosides would also fall into this category.

Cyanide species	Examples
Free cyanide	CN^-, HCN
Simple cyanide compounds (a) Readily soluble (b) Neutral insoluble	 NaCN, KCN, $Ca(CN)_2$, $Hg(CN)_2$ $Zn(CN)_2$, $Cd(CN)_2$, CuCN, $Ni(CN)_2$, AgCN
Weak complexes	$Zn(CN)_4^{2-}$, $Cd(CN)_3^-$, $Cd(CN)_4^{2-}$
Moderately strong complexes	$Cu(CN)_2^-$, $Cu(CN)_3^{2-}$, $Ni(CN)_4^{2-}$, $Ag(CN)_2^-$
Strong complexes	$Fe(CN)_6^{4-}$, $Fe(CN)_6^{3-}$, $Co(CN)_6^{4-}$, $Au(CN)_2^-$,

Cyanide toxicity decreases progressively down the table. The forms at the top of the table are the most toxic and those in the bottom line are virtually non-toxic. Hence environmental studies are most concerned with determination of cyanide species in the upper part of the table. The situation as regards cyanide in contaminated soil depends heavily upon the origin of the cyanide species and any reactions that may have taken place in the soil following deposition (cf. Kjeldsen 1999). For example, the soil from around iron oxide purifiers at MGP sites contains predominantly ferri and ferro cyanides and is of low toxicity (Shifrin *et al.* 1996). Conversely, some of the cyanide in gold mine tailings will exist in a number of metal species (generally as complexes depending on the nature of the co-mineralisation) and its potential toxicity will depend on the nature of the complexes (Smith & Mudder 1991).

Apart from these cyanide complexes, there are also mixed and bridged cyanide complexes (Barnes *et al.* 2000). A commonly occurring double salt is the Turnball or Prussian blue insoluble complex (Na or K)$Fe_4[Fe(CN)_6]_3$. In acidic soils, precipitation and dissolution of Prussian blue dominates iron cyanide speciation (Meeussen *et al.* 1994).

There are two general approaches to the determination of cyanide species:

1. Develop a generic (empirical) method whose digestion procedure releases cyanide only from the species of interest.
2. Undertake speciation studies, which will give exact concentrations of the forms of interest.

5.8.2 Thiocyanates

Comparatively little is known about thiocyanates, which contain the SCN ion (Boening & Chew 1999), particularly as regards their environmental fate and/or effects. The ion may arise through reaction of cyanide with sulfur (or sulfur compounds) under anaerobic conditions. In soil, thiocyanates are quite mobile compounds and in the absence of biological activity, show some resemblance in behaviour to chlorides. It is known that in soil, thiocyanate can be broken down into sulfide and HOCN, which is then hydrolysed to ammonia and carbon dioxide, although this is not the only possible breakdown pathway (Stratford *et al.* 1994). Thiocyanates are generally regarded as being of lower toxicity than the corresponding cyanides. They occur in some purifier wastes at MGP sites (Köster 2001) and also are common constituents of some gold mine tailings (Smith & Mudder 1991).

Thiocyanates have been measured in gold mine waste by colorimetry (Johnson *et al.* 2002), and in wastewater sludge (Kelada 1989). Both free cyanide and thiocyanate can be measured simultaneously with separation of cyanide from thiocyanate by quantitative conversion of thiocyanate to free cyanide with UV radiation (Meeussen *et al.* 1992). Thiocyanate concentration is calculated by difference.

5.8.3 Cyanates

Cyanates generally result from the oxidation of cyanides. The cyanate is generally an intermediate in the process and is slowly hydrolysed to ammonia and carbon dioxide. Little is known of the toxicity of cyanate, which is at least 100 times less toxic than free cyanide, and of similar toxicity to thiocyanate

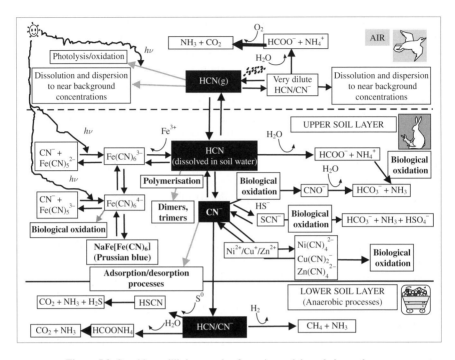

Figure 5.2 Cyanide equilibria, complex formation and degradation pathways.

(Environment Australia 1998). They are common constituents in waste streams at gold mines that use oxidative processes to destroy excess cyanide. Cyanate is commonly measured by:

1. Colorimetric techniques (Johnson *et al.* 2002) such as with 2-aminobenzoic acid (Guilloton & Karst 1985).
2. Reverse phase HPLC following conversion to a chromophoric thiocarbamyl derivative (Eiger & Black 1985).
3. Ion chromatography (Fagan & Haddad 1997).

While colorimetry is most commonly used, ion chromatography is the most likely future option.

5.8.4   Standard digestion procedures and free cyanide

Four procedures are available for the determination of cyanide in environmental media. These are: free cyanide, total cyanide, weak acid dissociable (WAD) cyanide and amenable to chlorination (ATC) cyanide. Free cyanide is that portion of the cyanide which exists as hydrogen cyanide or as the CN^- ion. Total cyanide is the cyanide released from a sample by digestion with strong mineral acid at elevated temperature. The cyanide released is distilled and trapped in an alkaline medium, where it is determined by an appropriate procedure. WAD cyanide is that fraction of the total cyanide that is liberated by digestion with weak organic acid (acetic acid). It is generally believed that this approximates the fraction of the total cyanide that is available to biota (i.e. that can be converted into the toxic species HCN at the pH of the gut). ATC cyanide is that fraction of the total cyanide that is destroyed by chlorination. It is most pertinent to the wastewater situation and has little application in the case of soils.

As regards determination of, free, total, WAD and ATC cyanides, there has been no major change to the standard digestion procedures employed in recent years. It is possible to extract free cyanide from soil by water extraction alone, and then apply a colorimetric or other analytical finish. However, although empirical, it is generally recognised that total WAD is the most practical form of cyanide to measure in soil samples. Free cyanide or other specific cyanide species are difficult to measure reproducibly, without the use of complex instrumentation. There has however been a number of attempts to improve the finish employed in general cyanide determinations. All three of the standard procedures involve a distillation step and this has traditionally been performed by classical methods. The existence of micro-distillation techniques and their obvious advantages have led to a number of trials as regards the determination of cyanide in soils. A number of micro-distillation techniques are available (e.g. Steig 1997). Mansfeldt and Biernath (2000) applied micro-distillation techniques to the determination of total cyanide in soils from former MGP and blast furnace sites. The results were found to conform closely with those of standard distillation procedures. Geetha and Balasubramanian (2001) described a spectrophotometric method using formaldehyde and 3-methyl-2-benzathiazolinone hydrazone for determining cyanide in the concentration range $0–0.2$ mg l^{-1}. The method was shown to be practically free of interference by thiocyanate and high concentrations of nitrate or nitrite. The possibility also exists for the use of instrumental methods as the finish to classical digestion procedures. For example, Okuno et al. (1979) have employed a gas chromatographic procedure for determination of cyanides in soil. Following acid digestion in an enclosed system, HCN was measured in the headspace with an electrical conductivity or a thermionic detector. (See also Appendix 1. ISO/FDIS 11262 and ISO/DIS 17380 for manual and automated easily liberatable and total cyanide methods.)

5.8.5 Speciation techniques for metallo-cyanide complexes

5.8.5.1 Ion chromatography
Pre-1991 work has demonstrated the usefulness of ion interaction chromatography in the separation of various metallo-cyanide complexes (Haddad & Kalambaheti 1991; Otu et al. 1996). Despite the overall applicability of

the method, earlier studies revealed certain problems. For example, the pre-concentration step gave quantitative recoveries of all species except $Fe(CN)_6^{4-}$ and $Cu(CN)_4^{3-}$. In the case of separation of Cu(I) cyanide, in extracts low in free cyanide, there was considerable fronting of the Cu(I) cyanide peak. Further studies have shown this to consist of a mixture of di- and tricyano complexes of Cu(I) (Fagan & Haddad 1997). The overall advantage of the technique is that it gives direct information on the content of the various metallo-cyanide complexes and thus permits a risk assessment of the particular soil to be carried out. Given the fact that cyanide determinations in soils have traditionally been performed by digestion procedures, a need exists for a comparison between this approach and that of ion interaction chromatography. In practise, such a comparison is virtually impossible, owing to the differing nature of the analyses (Huang *et al.* 1996).

5.8.5.2   Atomic absorption spectrometry

Atomic absorption spectrometry has been proposed as a method for the determination of free cyanide in aqueous solution and would also be applicable to soil extracts/digests. The principle of the method involves formation of a metallo-cyanide complex with the free cyanide (either contained in the sample or produced by a suitable digestion procedure) and then determination of the complexed metal by atomic absorption spectrometry. These methods are capable of detection limits as low as sub-ppb and are free from thiocyanate interference. Both gold (Blago 1989) and silver (Rosentreter & Skogerboe, 1991) have been employed as the complexing agents. Incorporation of selective oxidation, kinetic equilibrium and photodissociation techniques, allows speciation of cyanide compounds (Rosentreter & Skogerboe 1991).

5.8.5.3   Capillary zone electrophoresis

Capillary zone electrophoresis (CZE) provides an alternative to ion interaction chromatography for resolution of metallo cyanide complexes (Costa-Fernandez *et al.* 2000). The procedure may be interfaced with ICP-TOFMS, which permits further resolution of unresolved ionic species. The method is capable of detecting concentrations as low as $50 \mu g \, l^{-1}$.

5.8.6   Determination of other cyanide compounds in soil

As has been mentioned above, the possibility exists that soil may be contaminated by plant cyanoglycosides and from cyanide derived from this source. Combined enzymic digestion and microdiffusion procedures have been devised to permit determination of cyanoglycosides in soils and have been successfully applied to a number of field situations (Dartnall & Burns 1985).

5.9 Asbestos

5.9.1 Introduction – the asbestos minerals and their uses

Asbestos is the term for the fibrous form of a number of naturally occurring crystalline silicate minerals, which have been exploited commercially and are still mined and processed in various countries such as South Africa, Canada, the Former Soviet Union and China. The three main types of asbestos that have been commercially produced are Chrysotile, Amosite and Crocidolite, which are commonly referred to as White, Brown and Blue asbestos respectively. Other rarer forms of asbestos include Tremolite, Anthophyllite and Actinolite, which may be found either individually or as contamination within the three main types.

All of the above forms of asbestos belong to either the Serpentine or Amphibole mineral groups and have different chemical compositions and hence different associated properties. All have excellent heat insulating properties and are relatively incombustible, whilst some are flexible or have high tensile strengths, or resistance to acids and alkalis.

The properties of asbestos have been known for thousands of years, with one of the earliest recorded uses of asbestos being by Finnish Potters who used it to strengthen their clay over 4000 years ago. It is also recorded that the Egyptians and Romans shrouded their dead in asbestos cloth, and the Greeks and Romans made asbestos lamp wicks that apparently seemed to burn forever.

Due to its excellent heat insulating properties and fire resistance attributes, asbestos has been imported into the UK since 1880 and was widely used in building materials and products, with a peak usage in the 1960s. Due to the expansion of the housing stock and number of commercial buildings throughout the last century, a large demand and usage of asbestos containing materials (ACMs) occurred throughout the UK within the construction industry. This market was supplied by many UK factories, that processed the raw mineral into building materials or products and it has been estimated that around 70% of commercial buildings and housing stock built prior to 1975 would have originally contained some form of asbestos containing material (DETR 1999).

It is estimated that around six million tonnes of asbestos has been imported into the UK (DETR 1999), however the importation and use of Amosite and Crocidolite were prohibited in the UK in 1986 and Chrysotile was prohibited in November 1999, except for specialised applications such as high pressure steam gaskets etc. most of which will probably be discontinued by 2005. It should also be noted that the trade in, re-use or gifts of ACMs was also prohibited in 1999 and this includes asbestos cement roofing sheets that were previously often re-used when outbuildings were moved or altered.

5.9.2 The health effects of asbestos

Asbestos can split into microscopically small airborne fibres that are invisible to the human eye and are typically one to ten microns (one to ten thousandth of a millimetre) in length. They are typically two thousand times thinner than a human hair and can penetrate deep into the lungs and remain lodged virtually indefinitely within those that are exposed. The inert nature of these fibres means that the human body will not be able to break down the fibres and therefore they will cause irritation or damage to the delicate linings (pleural membranes) of the lungs for a long period of time after the initial exposure.

The health effects of long term or high levels of exposure to airborne asbestos fibres have been well documented since the early 1900s and were reported by the medical profession on many occasions in the first half of the last century. The majority of illnesses that are occurring at the present time are due to exposure forty to fifty years ago during the peak usage and manufacture of asbestos materials for the construction industry. These illnesses include the following fatal diseases:

Asbestosis. Scarring of the lungs, which results in the tissue becoming less elastic, the lungs become less efficient and therefore breathing becomes progressively more difficult.

Lung, throat and stomach cancer. These can be caused by all types of asbestos, however Chrysotile is considered to be less dangerous than the fibres of the Amphibole mineral group such as Amosite and Crocidolite.

Mesothelioma. Cancer of the inner lining of the chest or abdominal wall, which is almost exclusively caused by exposure to asbestos fibres.

5.9.3 Common building materials and products that contained asbestos

A surprisingly wide range of materials and products containing asbestos has been used within the construction industry that contained asbestos. In 1976, the asbestos industry was advertising the fact that asbestos was used in over 3000 different products (London Hazards Centre 1995). It is estimated that currently 4.4 million buildings still contain asbestos in the UK, of which 2.2 million buildings are domestic properties (HSC 2002). Examples of common building materials and products that contained asbestos are detailed below:

1. Loose pure asbestos fibre that was ready to be mixed on site as required, or to be used as a packing material around cables or placed into bags for insulation purposes etc.
2. Insulation materials such as sprayed coatings to concrete, reinforced steel joists etc. that were applied by spraying up to 85% asbestos mixed with a binder onto a sticky coating such as bitumen or adhesive glue.
3. Hard set insulation coatings that were usually mixed on site and hand applied/moulded onto pipework, calorifiers and tanks etc.

4. Sectional pipe insulation that could be quickly fitted and held in place with bands or string etc.
5. Rope, string, cloths, yarns, caulking and other woven gasket materials.
6. Paper, cardboard and backing to MMMF insulation or vinyl floor coverings.
7. Insulation boarding (AIB) that was often used for partitioning, ceiling tiles, door linings, firebreaks etc.
8. Millboard – a soft, low-density board with a high asbestos content used for fire protection.
9. Cement sheet and boarding materials such as roofing sheets, cladding, flat sheeting, rainwater services, water tanks, moulded decorative panels or containers, pipework/flue pipes, non-slip external floor tiles etc.
10. Gaskets, belts, brake shoes, clutches and other mechanical related materials.
11. Artificial slates, window sills, laboratory worktops.
12. Moulded plastic/resin materials such as toilet cisterns/lids, toilet seats, banisters.
13. Thermoplastic or vinyl floor tiles.
14. Bitumen roofing felt, damp proof courses, mastic sealant and adhesive type materials.
15. Textured decorative paint finishes for walls and ceilings.

At the present time, the majority of asbestos being mined is processed and manufactured into asbestos cement sheet or boarding materials. These materials are still in widespread use within the former Soviet Union, Africa, Asia, The Far East, Latin America and many of the Third World countries (various sources including *British Asbestos Newsletter*. This Newsletter publication is a very useful quarterly publication contributing something towards the understanding of asbestos-related topics in differing areas: legal, medical, historic, economic, corporate and sociological).

5.9.4 Common asbestos containing materials that may be found in contaminated land sites

In view of the vast range of possible asbestos materials or products that may be present within a building or throughout a site, it is not surprising that the demolition of buildings can result in ground contamination, especially considering that until recently, no great effort was made to diligently carry out thorough asbestos surveys and professional asbestos removal works prior to demolition or redevelopment.

Alternatively, industrial sites may have been granted a licence to bury waste materials on site and when pipework or plant was replaced, the insulation materials could have been crudely removed and disposed on site at a time when it was considered acceptable to carry out such practices. Often outbuildings and garages with asbestos cement roofs were demolished where they stood and all resulting materials used as hard standing for paths, roads or car parks etc.

Typical sites that may contain significant quantities of asbestos debris or ground contamination generally include railway land, depots, sidings, shipbuilders yards, dockyards, heavy engineering or industrial sites and waste disposal/landfill sites.

The common asbestos containing materials that may be found in contaminated land and brown field sites are therefore pieces of cement boarding, insulation boarding, fibrous insulation or occasionally free fibre depending upon the previous usage of the site, bitumen felt, vinyl floor tiles and gasket materials. These materials may have been mixed in with demolition rubble and soil, crushed as hardcore and utilised on site, or simply buried to raise the level of the land prior to redevelopment.

5.9.5   Encountering asbestos containing materials during a site investigation or redevelopment

If a site has a history of previous buildings being abandoned, demolished or any industrial processes that may have used asbestos, or led to waste materials being deposited on the site, then it is very likely that asbestos debris or materials will be present within the soil or fill material.

The potential presence of these materials can usually be identified by a desktop study, which typically would search through the documented history of the site and determine the previous usages and changes that it may have undergone. Additionally, a site investigation may determine the presence of building rubble or made ground that is obviously not natural and may contain unknown contaminants such as ACMs.

Prior to any works on site, a risk assessment should be carried out and possible control or preventative measures put in place to minimise any release of asbestos fibres and subsequent exposure of the operatives working on the site, or members of the public who may live, work or have right of way near the site boundary. If there is a strong possibility of ACMs being disturbed then it may be considered prudent to carry out reassurance air monitoring during the works to determine the airborne fibre levels and hopefully prove that no increase in the background fibre levels has occurred.

As with any work involving the disturbance of ACMs, the important fact to remember is that the airborne fibre release must be kept to an absolute minimum (as low as is reasonably possible) and that reliance upon RPE/PPE must not take precedence over the ability to reduce or minimise fibre release.

If a small amount of an ACM that is unlikely to release airborne fibres is uncovered, such as damp cement boarding that has been buried for many years, then it would be reasonable to put it aside, keep the material damp, place a temporary cover over the area and either backfill the material into the original area or arrange for the material to be removed from site by specialist contractors. However, if a large amount of an ACM that is highly friable and could potentially release a large quantity of asbestos fibres is uncovered, such as dry lagging materials, then work should be halted immediately, the material carefully dampened, temporarily covered prior to professional removal and advice sought as to whether the planned work should be stopped or continued under professional supervision and procedures amended so that the potential fibre release is minimised.

5.9.6 The packaging of suspected asbestos containing materials for transportation to a testing laboratory

If a material (including soil) that is suspected of containing asbestos is required to be positively identified to determine what type(s) of asbestos are present, then the material will have to be transported to a testing laboratory that contains specialised safety equipment such as a negative pressure safety cabinet, fitted with an *H Type* filter. The laboratory should preferably be UKAS accredited for such testing and would normally expect the following precautions to be taken prior to accepting any samples suspected of containing asbestos.

1. All samples should be double bagged and sealed in an airtight container such as polythene sample bags (up to 15 cm × 10 cm in size) or suitable plastic containers (up to 2 litres), within a heavy-duty plastic sack in the case of soil or fill samples.
2. Each sample should be clearly labelled in such a way, that the sample reference can be read without opening the sample and preferably with an appropriate hazard-warning label attached.
3. A chain of custody or brief letter detailing the client's instructions should accompany the samples with a clear note informing the laboratory that asbestos may be present within the sample bags or containers.

The remediation or removal of soil contaminated with asbestos from land under development is a very costly process. It is therefore essential that the laboratory be provided with samples that are of a suitable size, have been taken in a correct manner, stored correctly and are representative of the site or area from where they originated. To achieve these criteria, soil samples submitted for analysis should be in the range of 1–2 kg and be taken representatively throughout an area of potential contamination. In the case of large stockpiles awaiting classification prior to disposal or re-use, a guidance value of one sample per 250 m^3 or a minimum of four samples per stockpile is generally considered acceptable.

It should be noted that within the UK, the transportation of samples containing asbestos for analysis is only subject to the Carriage of Dangerous Substances Regulations, if the samples are contained in a receptacle of a capacity of 5 litres or more. However, it is not necessary to hold a waste transfer licence whilst transporting samples containing asbestos, but it should be noted that the testing laboratory must ensure that the final disposal of such samples is carried out in accordance with the UK Special Waste Regulations or relevant countries legislation.

5.9.7 Choosing a suitable testing laboratory

When choosing a suitable testing laboratory to carry out the analysis of bulk or soil type materials suspected of containing ACMs, the following requirements should be satisfied:

1. The laboratory staff should be suitably qualified and experienced to perform such work and must have received training regarding the hazards of working with asbestos.
2. The laboratory must be able to handle, analyse and repackage samples in a controlled environment where fibres are not released into the surrounding air or be vented out to the atmosphere.
3. The laboratory should be accredited for the type of testing requested and be independently audited on a regular basis. Suitable quality control requirements are enforced by the United Kingdom Accreditation Service (UKAS) and are required to comply with ISO 17025.

The laboratory staff carrying out the analysis should participate in external quality control proficiency schemes such as AIMS (the *Asbestos in Materials Scheme*, which is operated by the UK Health and Safety Laboratory) and they must be able to demonstrate sufficient history of obtaining satisfactory results.

5.9.8 Analysis of suspect asbestos containing materials

The analysis of materials that possibly contain asbestos can be carried out by various methods, however the dispersion staining microscopic technique is considered to be a quick, reliable and cost-effective method that is used by the majority of laboratories that routinely carry out this type of work. This method should be in accordance with MDHS 77 (1994), which is published by the UK Health and Safety Executive and forms part of the series of the methods for the determination of hazardous substances, titled *Asbestos in Bulk Materials* – sampling and identification by polarised light microscopy (PLM).

This method is generally applicable for the analysis of building materials that contain asbestos, however, if the material submitted for analysis comprises of a quantity of soil or fill type material, then careful screening will have to take place to remove all suspect ACMs from the soil. These suspect materials should be separated from the soil and washed if necessary prior to being analysed in accordance with MDHS 77 (1994) as detailed below.

The determination of the type(s) of asbestos present within a bulk material must be performed within a specialised safety cabinet fitted with a high efficiency HEPA filter and may be summarised as follows:

1. The bulk material could be prepared to facilitate analysis by various methods which may include cleaning, grinding, crushing, solvent washing, acid washing, ultrasonic treatment etc. These tasks must obviously be carried out in a manner that ensures no fibres are lost from the sample or released into the atmosphere.
2. The sample is initially examined under a zoom stereo microscope, individual fibres are carefully removed from the sample matrix and placed into an appropriate refractive index liquid to be examined.
3. The resulting slide is transferred to a microscope that is fitted with a dispersion staining objective and the fibres' characteristics such as morphology, colour, pleochroism, birefringence, extinction, sign of elongation, refractive index assessment and dispersion staining colours are examined.
4. The results of these observations are compared to reference materials and known characteristics, with all observations being recorded for quality control purposes and to provide a permanent record of the analytical findings, should the results be queried.

One of the problems associated with this analysis is that it is very difficult to accurately estimate the quantity (volume) of asbestos within a bulk material unless the asbestos fibres can be easily separated from the non-asbestos portion of the sample and their respective volumes compared. Routinely, an experienced analyst would be able to provide a visual semi-quantitative estimation, based upon their knowledge of the likely range of concentrations of asbestos within a certain type of material. This would be based upon their visual examination of the sample compared with the previous similar samples that they have analysed. However, this visual estimation can be subject to the following errors:

1. There may be non-asbestos fibres present which could look similar in morphology to asbestos fibres.
2. The material may not be homogenous or the sample submitted may be very small.
3. The sample submitted for analysis may be unrepresentative of the bulk material on site e.g. a surface scraping of a weathered cement product is likely to result in a high proportion of fibres compared to the actual cement product.

In view of the above difficulties, any estimation of the concentration of asbestos within a bulk material would therefore normally be reported as one of a series of ranges, such as trace (<2%), significant (2–50%) or substantial (>50%).

5.9.9   Achievable laboratory detection limits and interpretation of results

5.9.9.1   Bulk materials

When a bulk material is submitted for analysis, the laboratory should be able to locate and identify very small quantities of asbestos within the sample matrix. Polarised light microscopy methods should be able to detect 0.01% (100 mg/kg) asbestos if there are no interfering factors, however the majority of building materials or products contain concentrations ranging from virtually 100% asbestos down to around 1% and therefore it should not be difficult to correctly identify whether or not asbestos is present in the majority of routine samples.

If a material or product within a building is determined to contain asbestos then it will either need to be made safe, managed or removed depending upon the future planned usage of the building or site. In the UK, the Control of Asbestos at Work Regulations 2002 (CAWR), requires those with responsibilities for the repair and maintenance of non-domestic premises to determine if there are, or may be any ACMs present. This involves recording the location and condition of these materials together with an assessment and the management of any risk from them. Every non-domestic building (including the communal areas within tenanted domestic premises) will therefore be required to have an asbestos register or report available which can provide detailed information about the location and condition of known or potential ACMs and all relevant information must be issued to anyone who is potentially liable to disturb these materials before they commence their activities on site.

5.9.9.2   Soil materials

When a soil sample is screened for suspect ACMs, it should be possible to pick out pieces of boarding that are less than $0.1 \, \text{cm}^3$ in volume or small clumps of loose fibre providing that they are not totally encased in soil.

If these materials are weighed and were detected from within a 1 kg sample of soil, then it is possible to calculate (using density correction factors) a result that can be expressed as a percentage weight-to-weight of asbestos in soil. Using this method, it is possible to achieve a detection limit of between 0.001 and 0.01% w/w (10–100 mg/kg), depending upon the sample matrix.

In the case of soil or fill type materials that are submitted for analysis, there is no definitive guidance on what level of asbestos in soil might be considered safe or what usages may be permitted where low quantities of asbestos contamination are determined to be present.

If the soil material is due to be removed from site for disposal, then the waste disposal carrier and site accepting the waste will require a certificate of analysis determining how much asbestos is present. If the asbestos content is equal to or greater than 0.1% w/w (1000 mg/kg), then the material is classified in the UK as special waste (SI 972 1996) and can therefore only be disposed of at sites licensed to accept such waste. Any material containing less than 0.1% w/w would be generally classified as Category C waste (predominantly collected household waste, commercial waste and amenity waste, with other putrescible commercial and industrial waste) regardless of the quantity or type(s) of asbestos detected, down to a detection limit of 0.001% w/w.

To put these special waste limits in perspective, a 1 kg soil sample containing a single piece of cement boarding material weighing approximately 10 g and containing 12% v/v Chrysotile would be equal to the 0.1% w/w special waste limit. Clearly, if any significant quantity of cement boarding material is visible within a small soil sample that is due to be submitted for environmental analysis, then it could potentially be special waste material and should therefore be tested to determine the percentage of asbestos present.

At present, there is no official guidance defining what percentage of asbestos in soil would constitute a health risk and how this figure might depend upon the asbestos type, product, material, friability or moisture content etc. However, the UK HSE is also reviewing the test methods for asbestos materials in contaminated land and may provide additional guidance as part of this future documentation.

Abbreviations

ACM	Asbestos containing material
AIB	Asbestos insulation board
AIMS	Asbestos in materials scheme
DETR	Department of the Environment, Transport and the Regions
HEPA	High efficiency particulate arrester
HSC	Health and safety commission
HSE	Health and safety executive
MDHS	Methods for the determination of hazardous substances published by the Health and Safety Laboratory
MMMF	Man made mineral fibre
RPE	Respiratory protective equipment
PLM	Polarised light microscope
PPE	Personal protective equipment
UK	United Kingdom
UKAS	United Kingdom Accreditation Service

Notes

1. Use 1.0 g when expected S_{Cr} is in the range >0.5 to <1%. Use about 0.50 g when expected S_{Cr} is >1% and 3.0 g when <0.5%.
2. Dry soil immediately upon receipt at $85 \pm 5°C$ in a forced air draft oven, then grind to <0.5 mm particle size after removing shell, stones and litter.

References

Ahern, C.K., Baker, D.E. & Aitken, R.L. (1995) Models for relating pH measurements in water and calcium chloride for a wide range of pH, soil types and depths. *Plant and Soil*, **171**, 47–52.

AIMS (the Asbestos In Materials Scheme). This is run by UK Health and Safety Laboratory, Broad Lane, Sheffield, S3 7HQ, UK.

Aitken, R.L., Jeffrey, A.J. & Compton, B.L. (1987) Evaluation of selected extractants for boron in some Queensland soils. *Aust. J. Soil Res.*, **25**, 263–273.

Allison, J.D. & Brown, D.S. (1995) MINTEQA2/PROGEGA – A geochemical speciation model and interactive processor. In R.H. Loeppert *et al.* (ed.), Chemical equilibria and reaction models. *Soil Sci. Soc. Am. Spec. Pub. Madison*, WI. 241–252.

Alvarez, A. (1984) Standard solutions and certified reference materials. Ch 50. In S. Williams (ed.), Official Methods of Analysis of the Association of Official Analytical Chemists, 14th edn. (AOAC, Washington, USA.) pp. 1002–1004.

ANON (1990) Recommended practice for chemical analysis by ion chromatography. Australian Standard AS 3741. Standards Australia, Sydney.

Aomine, S. (1962) A review of research on redox potentials in paddy soils in Japan. *Soil Science*, **94**, 6–13.

APHA (1998) Part 4110. Determination of anions by ion chromatography. In Standard Methods for the Examination of Water and Wastewater, 16th edn. American Public Health Assoc., Washington, D.C.

Asbestos in Materials Proficiency Scheme: Administered by the UK Health & Safety Laboratory, Broad Lane, Sheffield, S3 7HQ, UK (Tel: **44 (0)114 289 2000).

ASTM (2000) D 4327-97 Standard test method for anions in water by chemically suppressed ion chromatography. In 2000 Annual Book of ASTM Standards, Section 11 Water and Environmental Technology, Vol. 11.01 Water (1) 422–427. ASTM, Philadelphia, PA.

Baker, D.E., Rayment, G.E. & Reid, R.E. (1983) Predictive relationships between pH and sodicity in soils of tropical Queensland. *Comm. Soil Sci. Pl. Anal.*, **14**, 1063–1073.

Barnes, D.E., Wright, P.J., Graham, S.M. & Jones-Watson, E.A. (2000) Techniques for the determination of cyanide in a process environment: a review. *Geostandards Newsletter*, **24**(2), 183–195.

Barrow, N.J. (1975) Reactions of fertilizer sulfate in soils. In K.D. McLachlan (ed.), Sulfur in Australasian Agriculture, pp. 50–57. Sydney Univ. Press, Australia.

Bates, R.G. (1962) Revised standard values for pH measurements from 0 to 90°C. *J. Res. Nat. Bur. Stds. – A. Phys. and Chem.*, **66A**, 179–184.

Berger, K.C. & Truog, E. (1939) Boron determination in soils and plants. *Ind. Eng. Chem. Anal. Edn.*, **11**, 540–545.

Blago, R.B. (1989) Indirect determination of free cyanide by atomic absorption spectroscopy. *At. Spectrosc.*, **10**, 74–76.

Boening, D.W. & Chew, C.M. (1999) A critical review: general toxicity and environmental fate of three aqueous cyanide ions and associated ligands. *Water Air and Soil Poll.*, **109**, 67–79.

Bohn, H.L. (1971) Redox potentials. [Soil samplings] *Soil Science*, **112**, 39–45.

British Asbestos Newsletter (http://www.lkaz.demon.co.uk/).

Brown, G. & Kanaris-Sotiriov, R. (1969) The determination of sulfur in soils by X-ray fluorescence spectroscopy. *The Analyst*, **94**, 782–786.

Canfield, D.E., Raiswell, R., Wetsrich, J.T., Reaves, C.M. & Berner, R.A. (1986) The use of chromium reduction in the analysis of reduced inorganic sulfur in sediments and shales. *Chem. Geol.*, **54**, 149–155.

Cartwright, B., Tiller, K.G., Zarcinas, B.A. & Spouncer, L.R. (1983) The chemical assessment of the boron status of soils. *Aust. J. Soil Res.*, **21**, 321–332.

CAWR (2002) The 'duty to manage asbestos' element of the CAW Regulations 2002, will come into force in Spring 2004. http://www.hse.gov.uk/asbestos/.

Chapelle, F.C., Haack, S.K., Adriaens, P., Henry, M.A. & Bradley, P.M. (1996) Comparison of Eh and H_2 measurements for delineating redox processes in a contaminated aquifer. *Environ. Sci. Technol.*, **30**, 3565–3569.

Conyers, M.K. & Davey, B.G. (1988) Observations on some routine methods for soil pH determination. *Soil Sci.*, **145**, 29–36.

Costa-Fernandez, J.M., Bings, N.H., Leach, A.M. & Hieftje, G.M. (2000) Rapid simultaneous multielemental speciation by capillary electrophoresis coupled to inductively coupled plasma time-of-flight mass spectrometry. *J. Anal. At. Spectrom.*, **15**, 1063–1067.

Dartnall, A.M. & Burns, R.G. (1985) A sensitive method for measuring cyanide and cyanogenic glucosides in sand culture and soil. *Biol. Fertil. Soils*, **5**, 141–147.

DETR (1999) Asbestos and man-made mineral fibres in buildings. (Aug 1999) ISBN 0–7277–2835–0.

Dick, W.A. & Tabatabai, M.A. (1979) Ion chromatographic determination of sulfate and nitrate in soils. *Soil Sci. Soc. Am. J.*, **43**, 899–904.

Donato, D. (1999) Bird usage patterns on Northern Territory mining water tailings and their management to reduce mortalities. Department of Mines and Energy Darwin. Northern Territory Government Printer, Darwin NT, pp. 1–36.

Eiger, S. & Black, S.D. (1985) Analysis of plasma cyanate as 2-nitro-5-thiocarbamylbenzoic acid by high-performance liquid chromatography. *Anal. Biochem.*, **146**, 321–326.

Environment Australia (1998) Cyanide Management (A booklet in the Series on Best Practice Environmental Management in Mining), Environment Australia Canberra.

Fagan, P.A. & Haddad, P.R. (1997) Reversed phase ion-interaction chromatography of Cu(I)-cyanide complexes. *J. Chromatogr.*, Series A, **770**, 165–174.

Farrell, R.E., Swerhone, G.D.W. & van Kessel, C. (1991) Construction and evaluation of a reference electrode assembly for use in monitoring *in situ* soil redox potentials. *Commun. In Soil Sci. Plant Anal.*, **22**, 1059–1068.

Fielders, M. & Perrott, K.W. (1966) The nature of allophane in soils. Part 3. Rapid field and laboratory test for allophane. *N.Z. J. Sci.*, **9**, 623–629.

Ford, H.W. & Calvert, D.V. (1970) A method for estimating the acid sulfate potential of Florida soils. *Soil Crop Sci. Soc. Florida Proc.*, **30**, 304–307.

Geetha, K. & Balasubramanian, N. (2001) Determination of cyanide by an indirect spectrophotometric method using formaldehyde and 3-methyl-2-benzathiazolinone hydrazone. *Anal. Lett.*, **34**, 2507–2519.

Gillman, G.P. & Bell, L.C. (1978) Soil solution studies on weathered soils from tropical North Queensland. *Aust. J. Soil Res.*, **16**, 67–77.

Guilloton, M. & Karst, F. (1985) A spectrophotometric determination of cyanate using reaction with 2-aminobenzoic acid. *Anal. Biochem.*, **149**, 291–295.

Haddad, P. & Kalambaheti, C. (1991) Advances in ion chromatography: speciation of mg L^{-1} levels of metallo-cyanides using ion interaction chromatography. *Anal. Chim. Acta.*, **250**, 21–36.

Hatcher, J.T. & Wilcox, L.V. (1950) Colorimetric determination of boron using carmine. *Analyt. Chem.*, **22**, 567–569.

Hayes, M.R. & Metcalfe, J. (1962) The boron curcumin complex in the determination of trace elements of boron. *The Analyst*, **87**, 956–969.

Hewitt, E.J. (1965) Ch. 20. In: Sand and Water Culture Methods Used in the Study of Plant Nutrition. Tech. Comm. No. 22, 2nd Edn. (Comm. Agric. Bur., England.)

HSC (2002) Consultative Document CD181 Issued February 2002 Published by HSE (Health and Safety Executive) Books (UK).

Huang, Q., Paull, B. & Haddad, P.R. (1996) Comparison of cyanide speciation by liquid chromatography and standard distillation methods. *Chem. Aust.*, **63**, 310–311.

James, B.R. & Bartlett, R.J. (2000) Redox phenomena. In Malcolm E. Sumner. (ed.), Handbook of Soil Science. CRC Press, Boca Raton, Fla. B169–B194.

Johnson, C.A., Leinz, R.W., Grimes, D.J. & Rye, R.O. (2002) Photochemical changes in cyanide speciation in drainage from a precious metal ore heap. *Environ. Sci. Technol.*, **36**, 840–845.

Kelada, N.P. (1989) Automated direct measurements of total cyanide species and thiocyanate, and their distribution in wastewater and sludge. *J. Water Pollut. Control Fed.*, **61**, 350–356.

Kjeldsen, P. (1999) Behaviour of cyanides in soil and groundwater: a review. *Water Air and Soil Poll.*, **115**, 279–307.

Köster, H.W. (2001) Risk assessment of historical soil contamination with cyanides; origin, potential human exposure and evaluation of intervention values. Bildhoven: National Institute of Public Health and the Environment.

Lindsay, W.L. (1979) Chemical equilibria in soils. Wiley-Interscience, New York, NY.

Liu, Z.-G. & Yu, T.-R. (1984) Depolarization of a platinum electrode in soils and its utilization for the measurement of redox potential. *J. Soil Sci.*, **35**, 469–479.

London Hazards Centre (1995) Asbestos Hazards Handbook (1995) ISBN 0–948974–13–3.

Mansfeldt, T. & Biernath, H. (2000) Determination of total cyanide in soils by micro-distillation. *Anal. Chim. Acta.*, **406**, 283–288.

Mansfeldt, T. & Biernath, T. (2001) Method comparison for the determination of total cyanide in deposited blast furnace sludge. *Anal. Chim. Acta.*, **435**, 377–384.

MDHS 77 (1994) Asbestos in bulk materials, sampling and identification by polarised light microscopy (PLM), June 1994, ISBN 0–7176–0677–5.

Meeussen, J.C.L., Keizer, J.C.L. & Lukassen, W.D. (1992) Determination of total and free cyanide in water (and soil) after distillation. *Analyst*, **117**, 1009–1012.

Meeussen, J.C.L., van der Zec, S.E.A.T.M., Bosma, W.J.P. & Keizer, M.G. (1994) Transport of complexed cyanide in soil. *Quad. – 1st Ric. Acque*, **96**, 5.1–5.5.

Morse, J.W. & Cornwell, J.C. (1987) Analysis and distribution of iron sulfide minerals in recent anoxic marine sediments. *Marine Chemistry*, **22**, 55–69.

Norrish, K. & Hutton, J.T. (1964) Preparation of samples for analysis by X-ray fluorescence spectrography. 1. Fusion in borate glass. 2. Powder samples. CSIRO Div. of Soils Div. Rep. No.3.

Norrish, K. & Chappell, B.W. (1967) In J. Zussman (ed.), Physical Methods of Determinative Mineralogy. Academic Press, New York.

O'Dell, J.W., Pfaff, J.D., Gales, M.E. & McKee, G.D. (1984) Test method. The determination of inorganic anions in water by ion chromatography – Method 300.0'. US Environmental Protection Agency, Cincinnati, OH.

Okuno, I., Whitehead, J.A., Higgins, W.H. & Savarie, P.J. (1979) A gas chromatographic method for determining residues of sodium cyanide in vegetation and soil. *Bull. Environ. Contamin. Toxicol.*, **22**, 386–390.

Otu, E.O., Byerley, J.J. & Robinson, C.W. (1996) Ion chromatography of cyanide and metal cyanide complexes: a review. *Int. J. Environ. Anal. Chem.*, **63**, 81–90.

Pang, H. & Zhang, T.C. (1998) Fabrication of redox potential microelectrodes for studies in vegetated soils or biofilm systems. *Environ. Sci. Tech.*, **32**, 3646–3652.

Ponnamperuma, F.N. (1972) The chemistry of submerged soils. *Advances in Agronomy*, **24**, 29–96.

Probert, M.E. & Samosir, S. (1983) Sulfur in non-flooded tropical soils. In G.J. Blair and A.R. Till (eds), Sulfur in South-East Asian and South Pacific Agriculture, pp. 15–27. Univ. of New England, Armidale.

Raiswell, R., Buckley, F., Berner, R.A. & Anderson, T.F. (1988) Degree of pyritization of iron as a paleoenvironmental indicator of bottom-water oxidation. *J. Sed. Petrology*, **58**, 812–819.

Raupach, M. & Tucker, B.M. (1959) The field determination of soil reaction. *J. Aust. Inst. Agric. Sci.*, **25**, 129–133.

Rayment, G.E. & Bruce, R.C. (1984) Soil testing and some soil test interpretations used by the Queensland DPI. QDPI Information Series QI 84029.

Rayment, G.E. & Higginson, F.R. (1992) Australian Laboratory Handbook of Soil and Water Chemical Methods. 330 Inkata Press, Melbourne.

Rayment, G.E., Walker, B. & Keerati-Kasikorn, P. (1983) Sulfur in the agriculture of Northern Australia. In G.J. Blair and A.R. Till (eds), Sulfur in South-East Asian and South Pacific Agriculture, pp. 228–250. Univ. of New England, Armidale.

Rosentreter, J.J. & Skogerboe, R.K. (1991) Trace determination and speciation of cyanide ion by atomic-absorption spectroscopy. *Anal. Chem.*, **63**, 682–688.

Shaikh, A.U., Hawk, R.M., Sims, R.A. & Scott, H.D. (1985) Graphite electrode for the measurement of redox potential and oxygen diffusion rate in soils. *Nuclear and Chemical Waste Management*, **5**, 237–243.

Shifrin, N.S., Beck, B.D., Gauthier, T.D., Chapnick, S.D. & Goodman, G. (1996) Chemistry, toxicology, and human health risk of cyanide compounds in soils at former manufactured gas plant sites. *Reg. Toxicol. Pharmacol.*, **23**, 106–116.

Siller, H. & Winter, J. (1998) Degradation of cyanide in agroindustrial or industrial wastewater in an acidification reactor or in a single-step methane reactor by bacteria enriched from soil and peels of cassava. *Appl. Microbiol. Biotechnol.*, **50**, 384–389.

SI 972 (1996) Statutory Instrument 1996 No. 972, (UK), The Special Waste Regulations 1996.

Smith, A. & Mudder, T. (1991) The chemistry and treatment of cyanidation wastes. London: Mining Journal Books Limited.

Steig, S. (1997) A miniature membrane tube for rapid parallel distillation of cyanide, phenolics, ammonia, sulfide, methylmercury and tritium from waters and solids. *Am. Environ. Lab.*, **9**, 10–11.

Stratford, J., Diaas, A.E.X.O. & Knowles, C.J. (1994) The utilization of thiocyanate by a heterotrophic bacterium: the degradative pathway involves formation of ammonia and tetrathionate. *Microbiology*, **140**, 2657–2662.

Stumm, W. & Morgan, J.J. (1996) Aquatic Chemistry. 3rd Edn. Wiley-Interscience, New York, NY.

Sullivan, L.A., Bush, R.T. & McConchie, D.M. (2000) A modified chromium-reducible sulfur method for reducing inorganic sulfur: optimum reaction time for acid sulfate soil. *Aust. J. Soil Res.*, **38**, 729–734.

Sullivan, L.A., Bush, R.T., McConchie, D., Lancaster, G., Haskins, P.G. & Clark, M.W. (1999) Comparison of peroxide oxidisable sulfur and chromium reducible sulfur methods for determination of reduced inorganic sulfur in soil. *Aust. J. Soil Res.*, **37**, 255–265.

Williams, C.H. (1975) The chemical nature of sulfur compounds in soils. In K.D. McLachlan, (ed.), Sulfur in Australasian Agriculture. pp. 21–30. Sydney Univ. Press, Australia.

Zhabina, N.N. & Volkov, I.L. (1978) A method of determination of various sulfur compounds in sea sediments and rocks. In W.E. Krumbein (ed.), Environmental Biogeochemistry: Methods, Metals, and Assessment. **3**, 735–745. Ann Arbor Science Publishers: Ann Arbor, MI.

6   Petroleum hydrocarbons and polyaromatic hydrocarbons

Jim Farrell-Jones

6.1   Introduction

Crude oil and the products derived from it are a complex mixture of compounds which change over time and distance when released into the environment. Added to this is the uncertainty related to definitions – for example, total petroleum hydrocarbons (TPH), often referred to as oil and grease, mineral oil, hydrocarbon oil and extractable hydrocarbons. Also, there are many analytical techniques available for measuring TPH concentrations. No single method measures the entire range of compounds covered by petroleum-derived hydrocarbons.

This chapter describes the chemical characteristics of petroleum fractions likely to be found in the environment and the factors that may influence the choice of analytical method. It also provides a review of analytical methods for total petroleum hydrocarbons (TPH), petrol range organics (PRO) and diesel range organics (DRO). Finally, there is an introduction to an analytical approach for a method of quantifying petroleum fractions, or individual constituents in conjunction with the TPH. This latter approach has been developed in conjunction with the total petroleum hydrocarbons criteria working group (TPHCWG) to assist in risk assessment at sites possibly contaminated by petroleum.

6.1.1   Chemical characteristics

Petroleum hydrocarbons are derived from crude oil which is predominantly a mixture of hydrocarbons and compounds containing heteroatoms such as nitrogen, sulfur or oxygen species (NSOs), together with low concentrations of metallo-porphyrins, carbon and hydrogen. Nearly all petroleum products are complex mixtures of a very large number of chemical components. Many are toxic, mobile and environmentally persistent. Regardless of the complexity, petroleum compounds are classified into two major component categories: hydrocarbons and non-hydrocarbons.

Hydrocarbons comprise the majority of components found in refined products and in most cases represent the species measured as TPH. These hydrocarbons can be grouped into classes such as aliphatic, aromatic and nitrogen, sulfur and oxygen compounds. Depending on the analytical method used, NSOs may be included with TPH measurements but are typically defined as non-hydrocarbons. Additionally, crude oils contain significant amounts of asphaltenes (high molecular weight resinous compounds, usually concentrated in the heavier distillation fractions and residues).

6.1.2 Architecture of organic molecules

Organic compounds account for more than 95% of chemicals in use today. Figure 6.1 shows how organic compounds can be divided into various groups, depending on their structure.

6.1.3 Aromatics

Aromatic hydrocarbons are of special commercial importance. The benzene ring structure, with six carbons and three double bonds, is the fundamental aromatic unit. This molecule can have one or more hydrogen substitutions with side chains, resulting in alkyl benzenes (e.g. the TEX in BTEX [benzene, toluene, ethylbenzene and xylene]) or two or more aromatic rings may be fused together to form polycyclic aromatic hydrocarbons (PAHs).

6.1.4 Chemical composition of petroleum

The main constituents of petroleum can be subdivided into five main groups:

1. naphthenes (cycloparaffins or cycloalkanes)
2. normal paraffins (normal alkanes)
3. iso-paraffins (iso-alkanes)
4. aromatics
5. nitrogen, sulfur, oxygen compounds (NSOs)

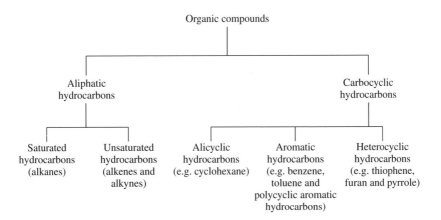

Figure 6.1 Classification of organic compounds.

(1) *Naphthenes*: The two central ring structures that predominate in crude oil are cyclopentane and cyclohexane, with five and six carbon atoms respectively.

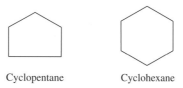

Cyclopentane   Cyclohexane

Many naphthenes contain the biogenic isoprene structure (see below) which can be used as biogenic markers for identifying petroleum origin.

(2) *Normal paraffins*: These straight chain structures are the simplest hydro-carbons. These compounds are readily biodegradable and occur as a continuous homologous series in most crude oils and form a major component. The boiling point increases regularly, with the first four in the homologous series (methane, ethane, propane and butane) being gases at ambient temperature, the next twelve are liquids (C5–C16) and above C17 are solids. They all have low chemical activity.

n-pentane

(3) *Iso-paraffins*: These are often associated with single carbon atom (methyl) branches. From C9 upwards, most iso-paraffins are isoprenoids (based on isoprene):

$$CH_2{=}C{-}CH{=}CH_2$$
$$|$$
$$CH_3$$

3-isopentane

Isoprene

Pristane and phytane are the most abundant isoprenoids in petroleum.

$$H_3C{\diagdown}$$
$$\quad\quad CH{-}CH_2{-}CH_2{-}CH_2{-}\overset{\displaystyle CH_3}{\underset{|}{CH}}{-}CH_2{-}CH_2{-}CH_2{-}\overset{\displaystyle CH_3}{\underset{|}{CH}}{-}CH_2{-}CH_2{-}CH_2{-}CH\overset{\diagup CH_3}{\diagdown CH_3}$$
$$H_3C{\diagup}$$

Pristane $C_{19}H_{40}$

(2,6,10,14-tetramethyl pentadecane)

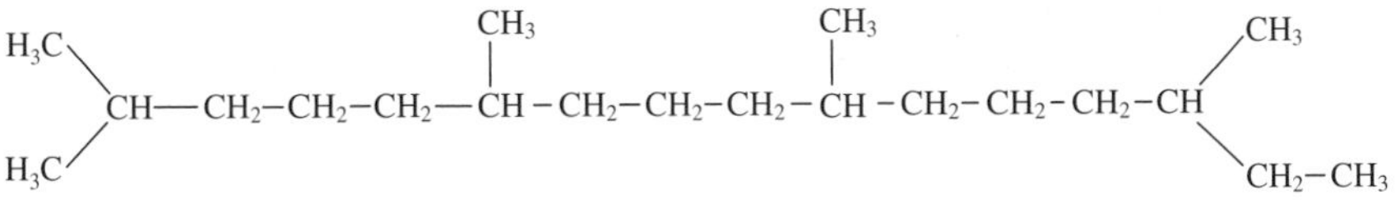

Phytane $C_{20}H_{42}$

(2,6,10,14-tetramethyl hexadecane)

(4) *Aromatics*: Originally meaning *fragrant* compounds, aromatics are now defined as 'Benzenes and those compounds which resemble benzene in their chemical behaviour'. Simple aromatics contain only one aromatic ring and may be either mono-, di- or tri-substituted, for example, one, two or three hydrogen atoms in benzene can be substituted to form methyl benzene (toluene), dimethyl benzene (xylene) and trimethyl benzene respectively as in Figure 6.2. Aromatic

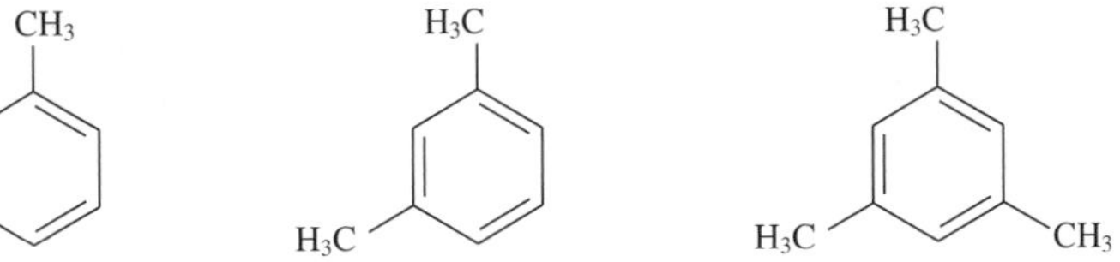

Methyl benzene   Dimethyl benzene   Trimethyl benzene

Figure 6.2 Mono-, di- and tri-substituted methyl benzenes.

compounds occur in crude oils and in most refined products as a series of isomers.

Multiple ring aromatic compounds are commonly referred to as polynuclear aromatic hydrocarbons or PAHs. The most common PAH is naphthalene. Naphthalene is a two-ringed unit. Three, four or more fused ring compounds also exist.

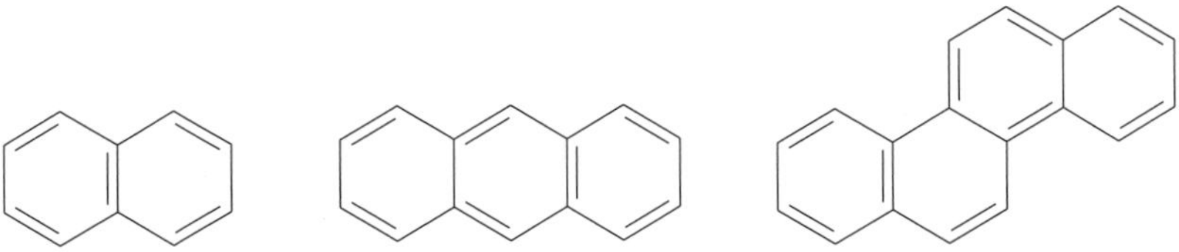

Naphthalene   Anthracene   Chrysene

(5) *Nitrogen, sulfur, oxygen (NSO) compounds*: These organic non-hydrocarbon compounds are found in crude oils and in various refined products. The sulfur heteroatom compounds such as thiophene and dibenzothiophene are found in motor spirits and diesel fuels. Other compounds such as pyrrole (nitrogen) and benzofuran (oxygen) are also prevalent in crude oils and its products.

6.1.4.1   Product characteristics

Refined petroleum products are primarily manufactured through distillation processes that separate fractions from crude oil by exploiting differences in their boiling point. Refinery operations are implemented to manufacture a product to a specification. These products vary from pale, mobile, and straw coloured liquids to highly viscous, semi-solid, black and tarry materials. The viscosities of such materials are highly variable and their densities vary from 0.79 to 0.95 g/cm^3. In addition, the manufacturing process may have increased the proportions of particular species (e.g. benzene), or decreased the levels of alkanes (depentanisation). A reduction of undesirable species such as sulfur (mercaptans) may have occurred and performance-enhancing additives would normally have been added.

Because of the complexity of these products, convention has dictated that these products are described by their boiling range. Because distillation cutoffs are broad and do not produce a sharp distinction (it is usual for a range to be quoted as being 95% boiling between 160 and 400°C), this invariably produces an overlap between fractions. The boiling point ranges correlate to the number of carbon atoms in the structure; the more the carbon atoms, the higher the boiling point. However, there are structural effects that also influence the boiling point. Branched and aromatic compounds often have boiling points that differ from their corresponding *n*-alkane analogs. It is for these reasons that boiling point defines the approximate carbon range. Generally with an increase in the size of a molecule, usually recorded as *carbon number*, there is an increase in both boiling and melting points, a decrease in vapor pressures and an increase in density. In addition, there is a general decrease in water solubility and stronger adhesion to soils resulting in less mobility in sub-surface conditions.

6.1.4.2   Gasoline (called petrol in the UK)

Gasoline, the blended product manufactured for combustion efficiency comprises more than 250 individual compounds found in the range C4–C12. Of this, more than 99% are found in the C5–C11 fraction with a boiling point range of 35–200°C. To increase yield and performance, refinery processes crack large molecules and alkylate or polymerise smaller molecules to produce molecules in the desired range. Octane ratings are enhanced by reforming or isomerising molecules that are in the desired range but burn unevenly. Oxygenates such as methyl tertiary butyl ether (MTBE) and alcohols are also added, typically as lead replacement antiknock additives. Other additives to enhance performance may also be added in small quantities. Until recently, alkyl lead additives such as tetraethyl lead (TEL) were added as octane improvers but have since been replaced by the lead-free replacement additives such as manganese tricarbonyl compound (MMT).

Gasoline comprises between 40 and 70% straight, branched and cyclic saturated compounds. Olefins are found in low concentrations mainly as a result of cracking larger molecules. Aromatic species range from 20 to 50% and include the common contaminants like benzene, toluene, ethyl benzene and xylenes (BTEX) and a small amount of naphthalene. These compounds are sufficiently soluble in water to leach into the groundwater and are mobile in the environment and relatively toxic. Fortunately, they are also amenable to natural attenuation processes.

Unleaded gasolines, in use today, comprise higher concentrations of BTEX compounds (Fig. 6.3) and branched chain hydrocarbons. Oxygenated additives such as methyl tertiary butyl ether (MTBE), tertiary amyl methyl ether (TAME), methanol and ethanol are also used as antiknock and combustion improvers.

6.1.4.3   Aviation gasoline

Traditionally aviation gasolines have a strictly limited composition designed to produce maximum octane ratings within a specific boiling range. Typically, they comprise 50–60% saturated hydrocarbons, 20–30% naphthenes and 10% aromatics, but olefins are usually absent. These leaded fuels are a blend of straight-run naphthas with iso-pentane and alkyl substituted benzene. Ethylene dibromide (EDB) is added as a lead scavenger. As with gasoline, the lead additive and scavangers are in the process of being replaced by the lead-free replacement oxygenates.

6.1.4.4   Jet fuels

There are two general grades of jet fuel:

- JP4 – a wide range heavy naphtha-kerosene blend. This aviation fuel is currently being phased out and replaced with an unleaded blend containing oxygenates and other alcohols as lead-free replacement additives. The carbon range covered is typically C6–C14.
- Jet A or A-1 or JP8 (US Air Force) – a kerosene used by the world's airlines. These fuels are essentially a fraction distilled from crude oil mixed with some cracked material. Jet A fuels consist of 70–90% saturated hydrocarbons, 10–20% aromatics, but up to 30% aromatics in kerosenes. Sulfur compounds and alkenes are removed by hydrotreating. Jet fuels, like kerosenes comprise hydrocarbons in the C8–C17 range but the majority are found in the C10–C14 range.

6.1.4.5   Kerosene – domestic heating fuel

Kerosene has essentially the same boiling range distillation as Jet A and is thus of similar composition. The main difference is the presence of sulfur species in kerosene. A low sulfur grade kerosene is produced as well as a regular high sulfur grade. The carbon range is C8–C16 and a typical composition is 70% saturated hydrocarbons, 30% aromatics. C10–C12 are the most abundant alkanes present.

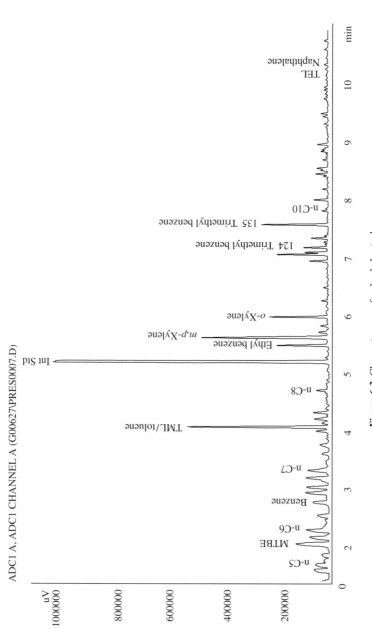

Figure 6.3 Chromatogram of unleaded petrol.

6.1.4.6   Diesel fuel

Diesel fuels are manufactured from distilled fractions of crude oil blended with cracked gasoils (gasoils which have been catalytically converted to produce smaller molecules that burn more efficiently). The major components of diesel are similar to crude oils, but include a higher aromatic fraction (10–40%). The typical carbon range for transportation diesel is C8–C28; however, the vast majority of constituents are found in C12–C20 range. Saturated hydrocarbons account for between 60 and 90% of diesel. The majority are naphthenes which appear as an unresolved envelope when analysed by gas chromatography. The biomarkers pristane and phytane are also dominant and used for forensic purposes. Diesel contains the targeted polycyclic aromatic hydrocarbons (USEPA target compounds). See Figure 6.4 for a typical *City* diesel.

6.1.4.7   Fuel oils

Fractions from crude oils that are heavier than diesel are often called residual fuel oils. These oils can be used as fuels or as feedstocks for other refinery processes to produce lighter and more valuable fuels.

The carbon number range covered is C20 to above C40. Fuel oils comprise saturated hydrocarbons, aromatic (15–45%) including PAHs and non-hydrocarbons (15–30%). These products are usually viscous, immobile and have a very low solubility in water.

6.1.4.8   Lubricating oils

Lubricating oils are high molecular weight, high boiling point fractions covering the range C26–C40 but in extreme cases the range is wider (C20–C50). Detailed compositions of lubricating oil fractions are difficult to obtain but it is known that these oils are usually enriched in cycloparaffins (naphthenes), aromatics and NSOs. The naphthenic envelope forms a characteristic profile, shown by gas chromatography, which changes as viscosity changes. The aromatic fractions predominantly consist of the fused ring type with phenanthrene derivatives more common among three-ring aromatics. NSOs tend to be concentrated in the high boiling fractions. These oils have extensive application in, for example, the automotive, aviation, marine and rail industries. Greases are lubricating oils to which a thickening agent such as graphite has been added. Detergents are also common additives.

6.2   Environmental fate of petroleum products

Petroleum products released into the environment undergo changes due to weathering effects such as evaporation, leaching, chemical oxidation and microbial degradation. The rate of weathering is highly dependent on environmental conditions. Leaching processes introduce hydrocarbons into the water

Current chromatogram (s)

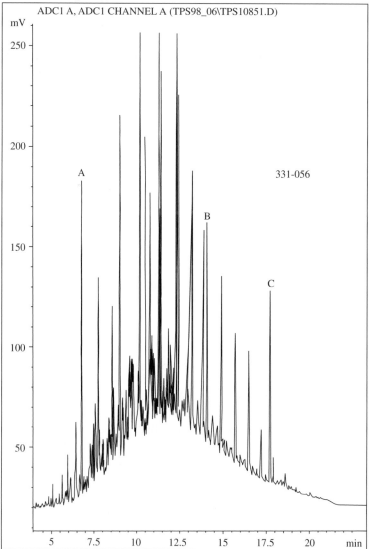

Figure 6.4 *City* diesel chromatogram.

phase by dissolution and entrainment. BTEX and MTBE tend to be the most water-soluble fractions and petroleum-contaminated groundwaters tend to be enriched with these substances relative to other petroleum constituents.

The rate of biodegradation is complex and dependent on external factors such as presence of microorganisms, nutrients and environmental conditions. However, the main factor governing the extent of biodegradation is the composition of each molecule. *n*-Alkanes are readily degraded while PAHs show variable degrees of degradation. The multiple ring cycloalkanes (naphthenes) show resistance. The isoprenoid compounds, pristane and phytane are almost non-degradable.

The above changes in composition may influence the test results and/or the method chosen to measure certain parameters. These potential changes need to be considered when specifying a test programme.

6.3   Factors influencing the analytical process

Before considering the analytical measurement, there are a number of common criteria that apply to all test methods.

- Collection and preservation.
- Extraction.
- Concentration.
- Clean-up.
- Measurement.

Each step affects the final result and should be included in any method-validation work.

6.3.1   Collection and preservation

Following the initial and critical step of designing a sampling plan, the implementation of that plan should ensure that a representative sample is collected. The sample must be stored and preserved to maintain the chemical and physical properties that it possessed at the time of collection.

The sample type, type of containers and their preparation, possible forms of contamination, and preservation methods must be considered to maintain the integrity of the sample. Risk assessors are encouraged to seek the advice of their analytical laboratory during the sampling design stage.

Additional problems are encountered with petroleum hydrocarbons due to the wide range in volatility, solubility and matrix effects. Knowledge of holding times should also be considered. Likewise, sample handling procedures within the laboratory will have a profound effect on the test result i.e. drying and crushing samples will lose volatile species, whereas a GC characterisation of a wet (*as received*) soil may sometimes show an unusually high aromatic content due to relative solubility effects.

6.3.2   Sample extraction

For most analyses, it is necessary to isolate the analytes of interest from the matrix (e.g. the soil, water). Extraction processes are quite complex and there is no universal procedure that caters for all analytes with the same degree of efficiency. However, for most petroleum fractions, the following methods have been used:

- Solvent extraction.
- Thermal extraction.
- Purging sample with inert gas.

Before considering the extraction technique, the choice of solvent should be assessed. The extraction of hydrocarbons from soils is largely influenced by the polarity of the solvent used. A change in polarity of a solvent will change the type and yield of the extract obtained (see also Chapter 8).

The solubility of organic compounds in selected solvents may be estimated from the *like dissolves like* rule of thumb. Whilst this approach is adequate, it is at best only qualitative and does not take into account the matrix and other effects present. Cocktails of solvents of different polarity can be prepared to give the same polarity as solvents which have or may be withdrawn from use for health and safety reasons. However, the presence of water in the sample can affect the polarity of the solvent and ultimately the extract. Soils can be oven dried up to 35°C with the significant risk of loss of volatiles or chemically dried (for instance, using magnesium sulfate monohydrate) to avoid volatile loss (see Chapter 8).

The choice of solvents is determined by cost, spectral qualities (for HPLC use), extraction efficiency, toxicity and commercial availability. Methylene chloride (dichloromethane) has been the preferred solvent for many semi-volatile compounds due to its high extraction efficiency and relatively low cost. However, for most petroleum species a non-polar solvent such as hexane is more effective for relatively fresh or recent spills. In aged polluted sites where absorption may have taken place, the addition of a polar solvent such as acetone to hexane is common. The hexane/acetone cocktail usually meets all requirements but may not always be compatible with the extraction technique. The following are examples of extraction techniques:

- Soxhlet.
- Manual shake.
- Accelerated solvent extraction.
- Microwave.

6.3.2.1 Soxhlet

Soxhlet is considered an exhaustive extraction and usually performed overnight. It is a continuous process which by its nature produces higher concentrations than other less exhaustive techniques and sometimes extracts unwanted substances. The process involves the soil sample being in contact with the solvent and with each cycle the extract liquid drains completely from the soil sample.

This process is repeated for up to 16 hours or until the extract liquid is clear. Water affects the polarity of the solvent and can also prevent contact between the extraction liquid and the matrix. It is therefore recommended that the soil be chemically dried or oven dried (up to 35°C with risk of loss of volatiles) prior to extraction. Crushing will also aid the extraction process.

6.3.2.2 *Manual shake*

This simple but effective procedure involves shaking the soil matrix with a suitable solvent for a period of time. It is suitable for most highly polluted soils (TPH of >1000 mg/kg). The lack of mechanical forces inhibits extraction when using non-polar solvents with moist soils. For this reason, acetone should be added to the non-polar solvent. The procedure can be modified to include ultrasonic agitation, which utilises high frequency sound waves to disaggregate the soil, thereby increasing the surface area of the sample in contact with the extracting solvent. Although this technique is widely used in standard methods for large molecules such as mineral oils, PCBs or PAHs, it is generally considered to be less efficient, but is relatively fast, cost-effective and often considered to be suitable as a screening test.

6.3.2.3 *Accelerated solvent extraction*

Accelerated solvent extraction is a closed system of extraction which utilises higher temperature and pressure. Closed systems are designed to minimise the loss of volatiles, improve the efficiency and increase the throughput. The elevated temperature improves analyte solubility. For example, anthracene, a PAH, is 15 times more soluble in methylene chloride at 150°C than at 50°C. High temperature also helps to overcome sample extraction matrix effects and gives faster desorption kinetics. The lower solvent viscosity allows diffusion of the solvent into the matrix to occur more quickly than other extraction techniques. Increased pressure also elevates the boiling point of the solvent.

Ideally, ASE works best when a dry, finely crushed and homogenised soil is used. Large clumps of soil and high moisture contents interfere with performance and can reduce extraction efficiency. The high moisture contents can be overcome by adding a water-soluble solvent such as methanol. The use of acetone is not recommended due to polymer-formation reactions occurring at high temperature and pressure resulting in erroneous peaks in subsequent chromatograms. Chemical drying is recommended when analysis of more volatile substances is required and for some tests, oven drying at 35°C may be used for semi-volatile species.

6.3.2.4 *Microwave*

As with ASE, microwave power can be used to extract contaminants from soil samples. The microwaves rapidly heat the solvents producing greater efficiencies in analyte recoveries and reduced extraction times. The solvents used need to be microwave active and samples must be free from metallic particles, which can limit the use of microwave extraction in soils.

6.3.2.5 Thermal extraction and static headspace

Thermal extraction techniques are usually performed in conjunction with gas chromatography. Petroleum hydrocarbons can be thermally desorbed from soil matrices at elevated temperatures. The eluting compounds are trapped in an absorbent phase such as Tenax™ and subsequently desorbed directly onto the column of the gas chromatograph. Whilst this technique is regarded as the closest to producing a real TPH value (C4–C35) it suffers from low sample size requirements (typically 1–10 mg) and is unlikely to be representative of the whole sample. Nevertheless, it can be used as a quick qualitative screening analysis.

The detection of low level concentrations of volatile petroleum hydro-carbons in either soil or water can be performed by static headspace analysis. In this technique, the gas phase in thermodynamic equilibrium with the matrix is analysed. The soil is placed in a headspace vial to which water and soluble salts such as sodium chloride are added to aid the transfer of hydrocarbons into the headspace. Internal standards and surrogate spikes can also be introduced. The vial is heated and an aliquot of the static headspace vapour is directly injected onto the column of the gas chromatograph. The advantages of this technique for volatiles such as gasoline range organics are less sample handling which minimises losses, no introduction of solvents which can inter-fere with the compounds of interest (MTBE), and the technique can be easily automated.

6.3.2.6 Purge and trap (dynamic headspace)

The purge and trap process is used when a lower detection limit (<10 µg/kg) is required and is considered a more efficient extraction process than static head-space. This is the basis of USEPA method 5030 'Purge and Trap for Aqueous Samples' and method 5035 'Closed System Purge and Trap and Extraction for Volatile Organics in Soil and Waste Samples' (see http://www.epa.gov/ epaoswer/hazwaste/test/main.htm). Method 5035 utilizes a hermetically sealed sample vial, the seal of which is not broken from the time of sampling to the time of analysis. Since the sample is never exposed to the atmosphere after sampling, the losses of VOCs during sample transport, handling and analysis should be negligible. The applicable concentration range of the low soil method is dependent on the determination method, matrix and compound. However, it will generally fall in the 0.5–200 µg/kg range. The vial is then placed, unopened, into the instrument carousel. Immediately prior to analysis, organic-free reagent water, surrogates and internal standards are automatically added without opening the sample vial. The vial is heated to ~40°C and the volatiles purged into an appropriate trap using an inert gas combined with agitation of the sample. Purged components travel via a transfer line to a trap. When purging is complete, the trap is heated and backflushed with helium to desorb the trapped sample components into a gas chromatograph for analysis by an appropriate determination technique.

Care should be taken when analysing soils by this process as the frits can easily be blocked by particles of soil. For higher levels (i.e. >1000 µg/kg) of contaminants in soils, the soil can be extracted prior to purge and trap analysis by shaking the soil in methanol and then taking an aliquot of this for analysis using method 5030, or the static headspace technique can be used. Method 5035 can be used for most VOCs with boiling points below 200°C and which are insoluble or slightly soluble in water. Volatile, water-soluble compounds can be included in this analytical technique. However, quantitation limits (by GC or GC/MS) are approximately ten times higher because of poor purging efficiency.

6.3.3   Concentration of sample extract

Concentration of the solvent extract prior to analysis allows for lower detection limits required to meet regulatory requirements. Concentration of extracts may be achieved by:

For volatiles (gasoline)

- by sorbant/cryogenic trapping.

For semi-volatiles (diesel)

- by Snyder column.
- by Kuderna Danish concentrator.
- by nitrogen evaporation.
- by vacuum evaporation.

6.3.3.1   Sorbant/cryogenic trapping

The sorbant trapping process in purge and trap analysis is essentially a concentration step, but combined with cryogenic focusing this affords better injection on to the chromatography column and loss of very volatile substances, e.g. vinyl chloride which may not be trapped on the absorbant.

6.3.3.2   Snyder column

Snyder columns are designed to allow highly volatile solvents to escape whilst retaining the semi-volatile compounds of interest. They are normally fitted onto the top of flasks containing extracts. The flasks may be heated, the design of the column allows volatile solvent to escape during the process. The analytes of interest condense from a vapour to a liquid phase and are collected in the solvent reservoir.

6.3.3.3   Kuderna-Danish concentrator (K-D concentrator)

The K-D concentrator is a Snyder column with a removable collection tube attached to the bottom. As the solvent is evaporated, the extract is collected in the collection tube.

6.3.3.4   Nitrogen evaporation

Sample extracts may be concentrated by nitrogen evaporation. Turbovap™ concentrators direct a steam of nitrogen over the extract surface. Vessels are housed in a temperature-controlled water bath. Although some losses of volatiles occur, this technique is suited to semi-volatile analysis such as DRO.

6.3.3.5   Vacuum

This, not very common method of evaporating solvents, can be used for higher boiling point fractions, but is unsuitable for volatile and semi-volatile extracts.

6.3.4   Clean-up of sample

Clean-up stages form an important part of many tests where interference by non-petroleum substances occurs. Invariably in soils, the solvent extracts non-petroleum substances such as humic acids and other naturally occurring organic compounds. These high molecular weight organic acids are polymeric compounds derived from the breakdown of plant and microbial materials. They can interfere with techniques such as infrared spectrometry and gas chromatography although with the latter, analysts can usually allow for these interfering substances.

The following clean-up steps, when required, can be used:

- Removal of non-petroleum species.
- Isolation of a particular species.
- Concentration of particular analytes.

6.3.4.1   Removal of non-petroleum species

Non-petroleum species such as animal and vegetable-derived hydrocarbons as well as naturally occurring hydrocarbons are commonly co-extracted from soils. These can interfere with the measurement technique and should, if possible, be removed prior to analysis. The most common technique is to pass the extract through a column containing a sorbent. The interfering compounds should have more affinity for the sorbent than the solvent and are retained on the sorbent phase. The compounds of interest should have no affinity for the sorbent and should pass through the column.

6.3.4.2   Isolation of particular species

The relationship between the polarity of solvent and the affinity for sorbent phases can be utilised in separating petroleum fractions into component classes. Most petroleum fractions contain aliphatics, aromatics and polar resins (NSOs), and these fractions can be isolated by column clean-up techniques. A glass column is prepared by adding two sorbent phases, silica and alumina. The extract, dissolved in a non-polar solvent such as hexane, is passed through the column. The aliphatic compounds have more affinity for the non-polar solvent than either of the sorbent phases and pass through the column and can be analysed as a separate fraction by GC or can be determined gravimetrically by weighing the extract after evaporation by one of the techniques highlighted above. The polarity of the solvent is changed and a known volume is passed through the column. Aromatic species previously retained on the silica have more affinity for the more polar solvent than the silica and pass through the column and can then be collected and treated as for the aliphatic fraction. The polar resins have more affinity for the alumina and are retained either indefinitely or can be eluted with a suitable very polar solvent.

The aliphatic and the aromatic fractions can be analysed together or separately. Confusion has arisen in the past with regard to terminology. Methods for the determination of mineral oil have been confused with that of TPH. To avoid confusion a definition that utilises the above clean-up as a principal component of the method is proposed: 'mineral oils are a group of compounds with the carbon number C10–C44, which are not retained when passed through a silica column using a non-polar solvent such as *n*-hexane'.

Chemically, mineral oils consist of aliphatics (*n*-alkanes, iso-alkanes and cycloalkanes). Interfering substances, such as elemental sulfur, can be dealt with by either passing the extract over activated copper, or for Soxhlet, the copper can be added at the extraction stage.

6.3.4.3   Concentration of particular analytes

Other petroleum species can be concentrated by molecular sieve clean-up which traps the long chain *n*-alkanes within the pore space of the sieves. Urea is another separation technique, based on molecular size. Urea crystals link together to form hexagonal chains enabling molecules of a certain size to be trapped or adducted into the crystal lattice, thus causing separation similar to molecular sieves. This technique is mostly used for removing *n*-alkanes thus concentrating the cyclic compounds. Acid/base partition can be used to separate the base/neutral and acid components by adjusting the pH.

As with all techniques there may be limitations to the use of clean-up techniques. Sample loading may exceed the capacity of the clean-up column. Some of the interfering compounds may have similar structures and their behaviour with regard to polarity and sorbent phases may be similar, thus evading clean-up. Some analytes of interest may be lost due to poor technique or choice of solvent or sorbent. An understanding of the chemistry of the analytes of interest and interferences may overcome some of the limitations.

6.3.5 Measurement

Once the sample preparations are complete, a number of techniques are available to quantify petroleum hydrocarbons:

- Total petroleum hydrocarbons (TPH).
- Petroleum group type analysis.
- Individual compound analysis.

6.3.5.1 Total petroleum hydrocarbons (TPH)
TPH has sometimes been referred to as mineral oil, hydrocarbon oil, extractable oil and oil and grease; this leads to confusion and misleading reporting. The definition of TPH depends on the analytical method employed and the concentrations of total compounds measured by that method. The same sample measured by different techniques may result in different concentrations being reported. Different methods are designed to measure different subsets or ranges of petroleum. No single method gives a precise and accurate measurement of TPH for all types of contamination. The main methods of analysis are gas chromatography, infrared spectrometry, gravimetric and immunoassay.

6.3.5.2 Petroleum group type analysis (not suitable for risk estimation)
Group type analyses are performed to measure the amounts of individual petroleum classes of hydrocarbons (e.g. saturates, aromatics, polars). This type of measurement is typically used for heavier fractions and can be used to interpret or identify the type of oil (Fig. 6.5).

Various methods can be employed, but the most common semi-quantitative method is a modified thin layer chromatography (TLC) technique. In addition, gravimetric group type analysis based upon column clean-up is used as the basis of speciated TPH (see later).

6.3.5.3 Individual compound analysis
For more detailed and generally more relevant analysis, individual compounds such as BTEX or individual PAHs can be determined by target compound analysis. This is generally performed by GC–MS. This precise identification and quantification is usually required for risk assessments.

6.4 Total petroleum hydrocarbons: a detailed method review

In general, a TPH method generates a single concentration based upon the total amount of petroleum compounds measured by a specific technique. The most frequently used methods are based on gas chromatography (GC), infrared (IR) or gravimetric analytical techniques. Of these, GC techniques are the preferred choice as they not only detect a broad range of hydrocarbons but also provide

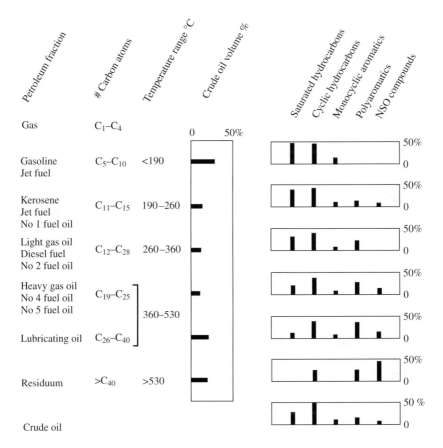

Figure 6.5 Group type chart.

better sensitivity and selectivity as well as identification. IR-based techniques have, in the past, been used extensively due to the fact that the test is quick, repeatable and inexpensive. However, it used the ozone-depleting solvent Freon 113. This method is non-specific and unable to provide information on the type of hydrocarbons present. Its use is therefore diminishing. Gravimetric techniques are both quick and inexpensive but are also non-specific. More recently, immunoassay TPH screening methods have been used, particularly for site monitoring of remediation progress or material classification.

6.4.1   Gas chromatography

For most GC-based methods, TPH is defined as any compound extractable by a solvent or gas and detectable by gas chromatography/flame ionisation detection

(GC–FID) within a specified range. The primary advantage of GC-based techniques is that they provide information on the type of hydrocarbons present. Quantitative interpretation of chromatograms requires an experienced analyst.

Gas chromatography is a technique for separating mixtures of compounds by partitioning the compounds between a flowing gas (mobile phase) and a non-volatile liquid phase (stationary phase). Separation is achieved by a combination of factors such as boiling point, polarity and compound affinity. Identification of compounds is achieved by measuring the time a compound takes to elute off the column. This *retention time* is characteristic of a compound under given criteria and column, and identification can be achieved by comparison with known substances.

Compound elution is generally in boiling point order. Eluted compounds are detected and the signal is proportional to their concentration. The sum of all responses within a specified range is equated to a hydrocarbon concentration with reference to standards of known concentration. Injection techniques include direct injection by syringe, headspace, or purge and trap.

A GC–FID will detect any hydrocarbon that elutes from a column and contains both carbon and hydrogen and burns in a hydrogen flame. GC-based methods are appropriate for detecting mainly non-polar compounds in the range C4–C35. Above C35, high temperature stationary phases are used. GC-methods can be modified to suit a particular range or suite of compounds such as GRO and DRO.

Interpretation of GC-based TPH can be complex and the analytical method used should always be considered when interpreting concentration data. A volatile range of TPHV can be used when gasoline is present as a pollutant but this method will not detect lubricating oils. Before attempting to interpret chromatograms, an understanding and knowledge of product types and ranges is essential. Gasoline is found in the volatile range whereas diesel is found in the extractable range, and between these is the semi-volatile range where jet fuels are found. However, there is an overlap between the volatile and extractable ranges (see Fig. 6.6).

Figure 6.6 acts as a guide to the typical ranges and tests associated with hydrocarbon analysis. However, the detection of different types of petroleum does not necessarily indicate multiple sources. Waste oils can contain a number of different products. Motor oil can contain both gasoline and diesel. Typically a used engine oil will show the characteristic diesel profile along with the lube oil pattern. Pattern recognition of fingerprints enables analysts to type products (see Fig. 6.7) but care must be taken as both weathering and biodegradation alter the fingerprint.

A fingerprint can be used to conclusively identify a mixture when a known sample of that mixture is available as a reference material. As most GC-based methods cover specific ranges, it is more appropriate to quote TPH measurement by range i.e. 'TPHD (total petroleum hydrocarbons diesel)'.

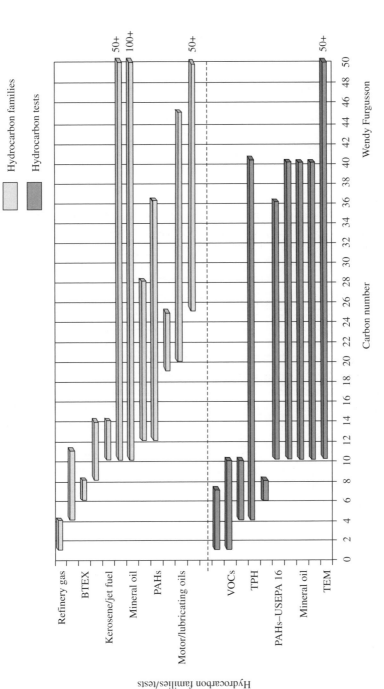

Figure 6.6 Hydrocarbon families and tests.

Current chromatogram (s)

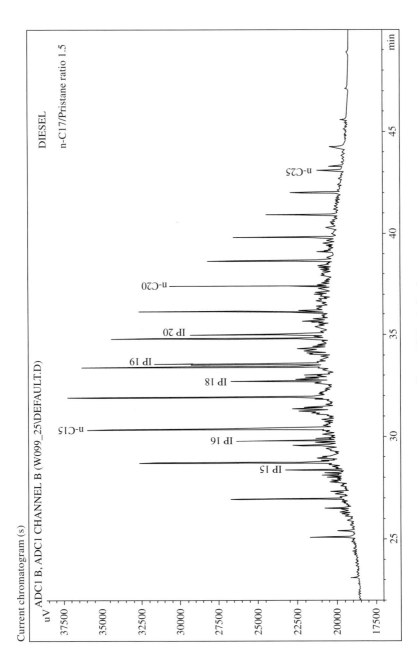

Figure 6.7 Fresh diesel.

Current chromatogram (s)

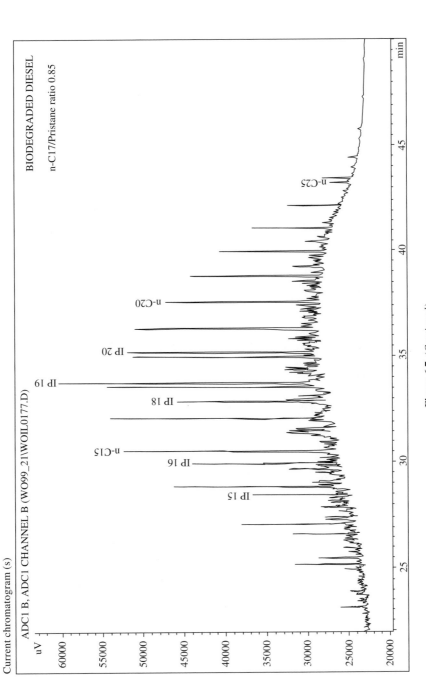

Figure 6.7 (*Continued*)

6.4.1.1   Gasoline range organics (GRO)

Gasoline (petrol) is a complex mixture of 250 individual compounds or more within a defined range of composition. It consists of saturates, olefins and aromatics together with oxygenated compounds such as MTBE. BTEX compounds are the most commonly analysed contaminants in this range. GRO analysis can be accomplished by using EPA methods 5000 series for the analysis of volatile organic pollutants. These methods employ either purge and trap or static headspace concentration with flame ionisation detection (FID) for the analysis of TPHV and BTEX.

The concentration step is used to extract the more volatile gasoline components from the matrix and is applicable to both soil and water. The method is used to determine the concentration of gasoline range (C4–C10). The range can be extended to cover up to C13, which includes naphthalene but this additional measurement should not be included in the GRO value.

With the advent of unleaded gasolines, the oxygenates, such as MTBE which have been used to replace lead additives, are readily soluble in water and have low vapour pressures making purge and trap less of an option especially when the recommended method of extraction of the soil by methanol is adopted.

Instead, headspace techniques can be adopted by which MTBE can be extracted, albeit at a low recovery. If response factors are applied, this can be a viable option. For water, headspace analysis can be conducted without any pre-treatment other than the addition of salts and internal standards. For soils, the sample should be slurried with pure water and salts, and internal standards can be added. After the concentration by dynamic headspace, an aliquot is injected onto a gas chromatograph equipped with flame ionisation detector. The area of the individual BTEX compound and MTBE peaks are determined separately and referenced to an internal standard (e.g. fluorobenzene). Calibration mixtures are also run to determine response factors relative to the internal standard. The concentration of the individual components can be determined. To determine the gross gasoline range, the total area, not including MTBE but including BTEX, can be summed and referenced against the internal standard (Fig. 6.8). A generic response factor determined by running a GRO standard mix is applied or a five point GRO calibration mix can be run and quantitation is performed by external calibration.

By using this approach the GRO concentration as well as BTEX and MTBE concentrations are determined. TPHV (C4–C13) can also be quoted especially if aviation kerosenes are apparent. Tetraethyl lead can also be detected.

6.4.1.2   Diesel range organics

Diesel range organics (DRO) methods of analysis are designed to measure the concentration of mid-range petroleum products such as diesel fuel, domestic heating fuels and aviation fuels in the range C10–C28. However to cover a wider variety of products, this range can be extended to C44. It is applicable to both

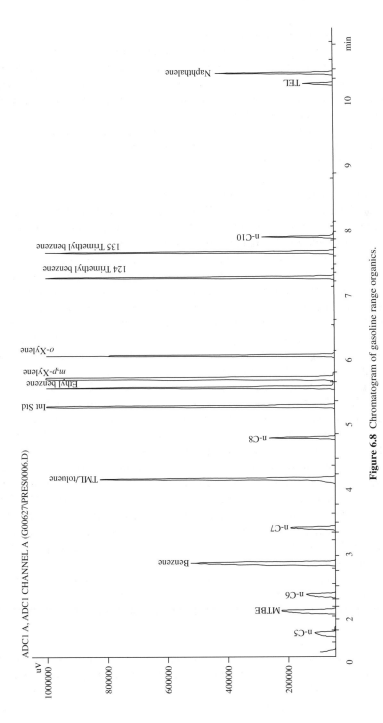

Figure 6.8 Chromatogram of gasoline range organics.

waters and soils. To determine the TPH value the whole chromatogram is quantified. But individual ranges can be quantified separately if required. With most products in this range, a characteristic envelope (Unresolved Complex Matrix) is apparent. This should be integrated with all other peaks in the chromatogram. Quantitation can be done by external calibration whereby calibration mixtures of known products are prepared and a calibration graph is produced, or by internal standardisation. In this procedure, three standards are used: Chlorooctadecane, heptmethylnonane and squalane are used. The latter two serve as surrogate standards put in prior to the extraction stage to assess recovery, the former is added after extraction and is used as an internal standard. This method not only quantifies all hydrocarbons in the range (care should be taken to correct non-hydrocarbons such as humic acids that commonly elute after C28) but also provides information on product type, degree of weathering and biodegradation. Ageing of diesel hydrocarbons and fractions that contain isoprenoids can also be estimated by experienced analysts. Figure 6.9 shows a weathered diesel where the alkanes have been lost due to evaporation. This is in contrast to Figure 6.7 of a fresh diesel. The method can also be used to assess mineral oil solely by analysing the fraction after clean-up by silica gel or indeed the PAH fraction after clean-up through silica/alumina.

6.4.1.3 Total petroleum hydrocarbons

As no specific standard methods cover the full range C4–C44 for TPH, a combination of GRO and DRO can be used to assess TPH. The concentration of C4–C10 from the GRO method and that of C12–C44 from the DRO method can be summed. The overlap in ranges of products can be accounted for by averaging the concentration of C10+ determined in the GRO method with the C12– in the DRO method. These two values should correlate within ±10% and this average value can be added to the GRO and DRO to provide a TPH value. Simply adding the GRO and DRO measurements is only a guide but not a definitive measure of TPH.

A recent development in GC technology is the use of the EZ flash that uses fast temperature programming with conventional GC ovens. This fast temperature programming using resistive heating in conjunction with special columns allows high boiling point mixtures to be analysed in seconds rather than minutes. There is some loss in resolution which may impair interpretation but, in general, this technique can provide a rapid TPH in the range C8–C40. Samples are extracted and directly injected onto the column of the gas chromatograph. Quantitation is normally done by external calibration. This is an exciting prospect for the future and is being validated in many laboratories.

Current chromatogram (s)

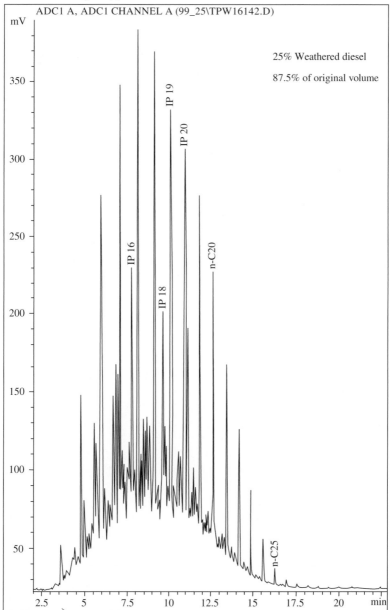

Figure 6.9 Weathered diesel.

Current chromatogram (s)

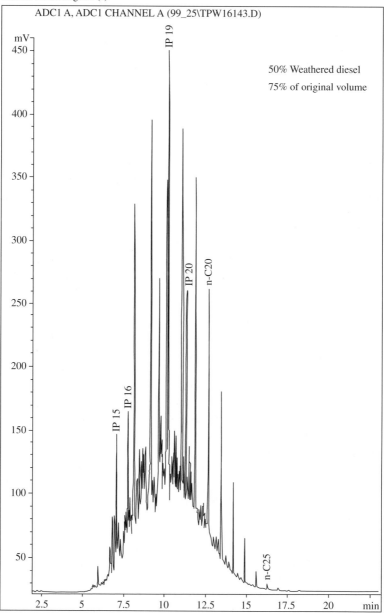

Figure 6.9 (*Continued*)

6.4.2   Infrared spectroscopy (IR) TPH

For IR-based methods, TPH is defined as any compound extractable by a solvent (Freon 113 or tetrachloroethylene), which is not removed by silica gel or florisil clean-up and which can be measured at a specified wavelength. This definition is similar to that of mineral oil discussed previously. The method most referred to is the EPA Method 418.1 which is used for recoverable petroleum hydrocarbons. This method often suffers from poor accuracy and precision, especially for heterogeneous soil samples. This method gives neither the information on petroleum type nor the presence or absence of toxic materials, and is thus of little use in risk estimation.

IR-based methods measure the vibration (stretching and bending) that occurs when a molecule absorbs energy in the infrared region of the electromagnetic spectrum. Different functional groups and bond types have different IR absorption frequencies and intensities.

The IR-based TPH Method 418.1 measures the absorbance of the C–H bond at a frequency of $2930\,cm^{-1}$, which corresponds to the stretching of aliphatic CH_3 group. In biodegraded fuels, n-alkanes are normally absent, thus making this method even less reliable. For biodegraded oils, it is more prudent to use three bands such as $2960\,cm^{-1}$ for CH_3 groups, $2930\,cm^{-1}$ for CH_2 and $3030\,cm^{-1}$ for aromatic C–H bonds.

Soil samples are extracted using tetrachloroethylene, a solvent which does not contribute any C–H stretching to the measurement. Others such as Freon 113 were commonly used, but this has been withdrawn and replaced in many laboratories by tetrachloroethylene. Quantitation is performed by comparison with known standards. It is important to use a calibration standard similar to the contaminants in the sample. Calibration using fresh diesel or kerosene will produce erroneous results if the contaminants are biodegraded. EPA Method 418.1 specifies the use of a n-hexadecane : isooctane : chlorobenzene mixture. Use of a dissimilar standard will tend to create a positive bias (i.e. overestimate) in highly aliphatic samples and a negative (i.e. underestimate) in highly aromatic samples. IR-based methods are also prone to interferences from non-petroleum sources, as most organic compounds have some type of alkyl group associated with them.

In general, results obtained from Method 418.1 should not be expected to match the extractable range of GC-based results. IR-based methods are not limited to carbon range and will detect heavier hydrocarbons than GC-based methods. Responses for different groups are variable and single wavelength measuring at $2930\,cm^{-1}$ will not detect benzene or naphthalene because of the absence of alkyl groups. Different ways of quantifying the absorbance maximum at $2930\,cm^{-1}$ can lead to significant variations in results, if only C–H absorbance is measured. IR-based methods will underestimate TPH concentrations.

6.4.3   Gravimetric TPH methods

Gravimetric methods measure any solvent extractable compounds that are not removed by the evaporation stage and are capable of being weighed. If this method is used as a measure of TPH, a clean-up to remove polar material is required, otherwise the data could not be considered as a valid TPH measurement, but can be used as an oil and grease measurement. Gravimetric methods give no information on hydrocarbon type. Gravimetric clean-up can be performed to give information on classes of compounds i.e. saturates, aromatics, polar substances. All results should be reported on a soil dry-weight basis and as such are only suitable for heavier fractions. As with all methods, the choice of solvent will influence the result but typically for TPH measurements hexane is recommended if no clean-up is involved.

6.4.4   Immunoassay TPH methods

Immunoassay methods correlate TPH with the response of antibodies to specific petroleum compounds. Various test kits based on immunoassay are available as self-contained systems that are useful for on-site measurement. These techniques are considered screening methods as specificity, precision and accuracy are lower than GC-based or IR-based methods.

Immunoassay techniques rely upon synthetic antibodies that have been developed to form a complex with petroleum substances. The antibodies in the test kit are immobilised on the walls of a special cell or membrane. Water samples can be added directly, whereas soils are solvent extracted into a suitable water miscible solvent and added to the cell. A known amount of enzyme with an affinity for the antibody is added. After equilibrium is established, the cell is washed to remove any unreacted material. Colour development reagents which react with the enzyme are added. A solution that stops colour development is also added at a specific time, and the optical density is then measured. Samples showing high optical density (colour intensity) contain low concentrations of analytes. Concentration is inversely proportional to optical density. Kits are generally available for, among others, TPH, BTEX and PAH. A correction factor supplied by the manufacturer is used to calculate TPH and this is subject to variation depending on the product type. These tests do not provide information on product type and have limitations dependent upon soil type and homogeneity. Also, field extraction techniques are not as efficient as laboratory-based extraction techniques.

6.5   Petroleum group type analysis (detailed review)

Group type analysis methods are designed to separate hydrocarbon mixtures into classes such as saturates, aromatics and polars. As described previously, the simplest form of class separation is a modified gravimetric clean-up using silica/alumina. Other methods have been used (mainly at refineries) such as HPLC. This utilises refractive index (RI) and ultraviolet (UV) detection typically at 254 nm and silica columns to separate hydrocarbon mixtures into the compound classes. A modified TLC approach using flame ionisation detection overcomes the shortfall in HPLC detectors with respect to variable response of PAHs with fixed wavelength UV detectors.

6.5.1   Thin layer chromatography (Iatroscan™ method)

This modified method is used as a screening method to separate the main components of petroleum into classes. The soil sample is extracted in a polar solvent such as methylene chloride and carefully evaporated to dryness using nitrogen (Turbovap™). The extract is weighed and a non-volatile hydrocarbon content is determined. The extract is dissolved in a non-polar solvent such as heptane and an aliquot is injected into previously activated alumina rods. The rods (ten to a frame) are placed in a bath containing a non-polar solvent (heptane) and by capillary action the solvent moves up the rods along with the saturated hydrocarbons which have no affinity for the alumina. At a given time, determined by the aromatic band starting to move up the rod, the rods are taken out of the bath. The rods are dried and the origin ends placed in a bath containing a higher polarity mix. The aromatics, which have more affinity for this solvent than the alumina, transpose up the rods, again at a given time the chromatography is curtailed. The solvent is replaced with a very polar solvent and the process is repeated and the NSOs move up the rods. This analysis can also be further extended to cover asphaltenes. The rods are placed on an IATROSCAN® scanner which has a flame ionisation detector. The rods are passed through the flame and the bands cause ionisation similar to GC–FID. The area in response is proportional to concentration, a standard oil of known composition is run with every batch of ten samples. The percentage composition is calculated by summing the areas of each peak, and a response factor is applied to each class. The concentration of the solvent extract is used to determine the concentration of each class present in the extract. Although this technique does not provide information on product type, the ratio of saturated compounds to aromatic compounds to asphaltenes is a useful parameter (see Fig. 6.10).

6.5.2   Speciated group type TPH

New methods of presenting TPH data that quantify the TPH as a series of ranges instead of a single measurement have been developed. These data can be used to assess risk. TPH fraction methods can be used to measure both volatile and extractable hydrocarbons in the range C4–C35. The ranges chosen can be modified to suit regulatory requirements. The basis of the method is to report

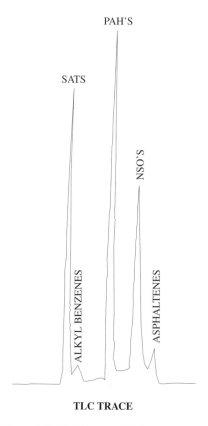

TLC TRACE

Figure 6.10 TLC (Iatroscan™) chromatogram.

separate concentrations for discrete aliphatic and aromatic fractions. Methods are GC-based and are similar to the GRO and DRO.

The soil sample is either extracted in *n*-pentane or a combination of head-space and extractable methods similar to GRO and DRO is used. The latter has already been discussed earlier. For the *n*-pentane extract, the whole fraction is separated by column chromatography as previously described. Care should be taken on the change over from aliphatics to aromatics. This is best achieved by using fluorescence to indicate when the aromatic band is approaching elution from the column. Before attempting column chromatography, the technique should be standardised by using calibration mixtures to perfect the cut-off and to ensure that PAHs do not elute in the aliphatic range. The fractions are separated and analysed by GC and specified ranges quantified similar to GRO and DRO. This approach not only provides more information used for risk assessment but also provides individual chromatograms of mineral oil, total PAHs as well as individual BTEX compounds. Speciation of PAHs can be undertaken.

6.6   Individual compound analysis

GC–MS is used to identify individual hydrocarbon compounds. Extraction and introduction of sample on to the GC–MS are similar to that of GC. GC–MS usually provides confirmation of the identity of the analyte through retention time and unique mass spectral pattern. The advantage of GC–MS is the high selectivity or the ability to confirm identity. Mass spectrometers are designed to ionise compounds and to scan for ions of specific mass-to-charge ratios. Each compound fragments into a consistent pattern of fragment ions. Every compound with the exception of some isomers has a unique mass spectrum. A plot of a single ion over time is called an ion chromatogram. These chromatograms can be used to identify specific types of compounds in the presence of complex mixtures. The total ion chromatogram is similar to a GC chromatogram. GC–MS can be set up to monitor specific compounds such as PAHs or BTEX by a process called selective ion monitoring.

6.7   Polyaromatic hydrocarbons (PAHs)

6.7.1   Structure

Sometimes referred to as PNAs (polynuclear aromatics), these are ubiquitous compounds found in many sites contaminated with petroleum or coal residues. They consist of two or more fused benzene rings. The PAHs of most interest in environmental chemistry are the 16 USEPA PAHs. These compounds were selected as representative, after analysis of numerous samples in the USA found that over 80% of the total PAH contribution could be attributed to these 16 priority compounds. The full list (with the number of rings in brackets) is as follows (see Fig. 6.11).

Further studies performed in the UK have led to the suggestion that three additional PAHs, anthanthrene, benzo(e)pyrene, cyclopenta(c,d)pyrene, often found associated with waste from coal carbonisation (town gas) plants, should also be included.

6.7.2   Sources

PAHs occur in petroleum products such as kerosene, jet fuel, fuel oils, diesel, engine oils and petrochemical waste. The other major source is from coal derived products, particularly manufacturing gas plants, cooking operations and wood preserving sites. PAHs are introduced into the environment as a product of natural and incomplete combustion of fossil fuels.

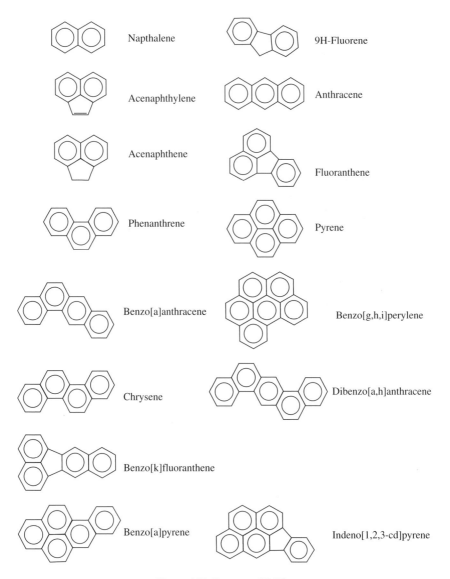

Figure 6.11 Structure of PAHs.

6.7.3 Carcinogenic nature

One of the most significant and harmful effects of PAHs is due to their carcinogenic nature, although this is not consistent across the whole range of compounds. Eight of the four, five and six membered ring PAHs are considered to be carcinogens.

PAH	Number of aromatic rings
Benzo(a)pyrene	5
Dibenzo(a,h)anthracene	5
Indeno(1,2,3-cd)pyrene	5
Benzo(b)fluoranthene	4
Benzo(k)fluoranthene	4
Benzo(a)anthracene	4
Chrysene	4
Benzo(g,h,i)perylene	6

Benzo(a)pyrene and dibenzo(a,h)anthracene are the most potent among these eight carcinogens.

Several studies have proposed the use of toxicity equivalent factors (TEF) being applied to PAHs, in order to accurately reflect the risks posed by specific sites and the distribution of PAHs present. This involves benzo(a)pyrene being given a TEF of 1, and all other PAHs given a TEF in accordance with the degree of carcinogenicity in relation to this compound, e.g. naphthalene has a TEF of 0.001, as it is considered to be a thousand times less carcinogenic than benzo(a)pyrene, and chrysene has a TEF of 0.1. These factors are then multiplied by the actual individual PAH concentrations in mg/kg and summed to give a total TEF for the sample expressed as BaP (benzo(a)pyrene) equivalents. This approach is routinely used for dioxin/furan analysis.

Compound	Toxicity equivalent factor
Naphthalene	0.001
Acenaphthene	0.001
Acenaphthylene	0.001
Anthracene	0.001
Benzo(a)anthracene	0.1
Benzo(g,h,i)perylene	0.1
Benzo(b)fluoranthene	0.1
Benzo(k)fluoranthene	0.1
Benzo(j)fluoranthene	0.1
Benzo(a)pyrene	1.0
Benzo(e)pyrene	0.001
Chrysene	0.1
Dibenzo(a,h)anthracene	1.0
Fluoranthene	0.001
Fluorene	0.001
Indeno(1,2,3-cd)pyrene	0.1
Phenanthrene	0.001
Pyrene	0.001

The adoption of the Index Dose as the health criteria value for non-threshold (*carcinogenic*) substances is probably going to undermine the basis of the TEF concept (see Appendix 3; CLR 9 and Tox 2 for details). Recent thinking also suggests that the above TEF values may not be accurate.

6.7.4 Methods of analysis

There are several methods for analysing PAHs, depending upon the level of accuracy, detection limit and specificity required:

- Screening test kits – enzyme linked immuno-sorbent assay (ELISA)
- Gravimetric (semi-quantitative)
- Thin layer chromatography (TLC)
- High pressure liquid chromatography (HPLC)
- Gas chromatography–flame ionisation detection (GC–FID)
- Gas chromatography–mass spectrometry (GC–MS)

The first three methods give a total PAH concentration. The last three methods can give all relevant individual PAH concentrations and are the most commonly used methods.

6.7.4.1 Screening test kits

The ELISA kits are based on the use of selective antibodies raised against specific PAHs, which are attached to a solid matrix support. These are combined with sensitive enzyme reactions (see Section 6.4.4). The kits provide high selectivity and can provide an immediate answer on site, but the disadvantages are that they can be difficult to use in the less than ideal conditions on site and they can only usually give a range of concentrations, e.g. >50, 20–50, 1–20 or <1 mg/kg of total PAHs and mainly rely on a manual rapid cold shake extraction technique which may only extract a small fraction of the PAHs in some samples. In addition, the test kits can prove quite expensive for a screening analysis.

6.7.4.2 Gravimetric

Solvent extractions of known concentration are passed through a silica or Florosil type chromatography column, and separated into the three major classes of compounds – aliphatics, aromatics and asphaltene/polars. Each fraction is eluted using a different polarity solvent, and the fraction is collected in a pre-weighed vial. This is evaporated to dryness and re-weighed to produce a result expressed as mg/kg of total PAHs calculated from the original solvent extract and weight of soil extracted. The result is more accurate than the kits, down to a detection limit of ~50 mg/kg. This semi-quantitative method is seldom used by contract laboratories. It also requires large volumes of organic solvents.

6.7.4.3 Thin layer chromatography (TLC)

This method uses an instrument known as an Iatroscan™, and involves spotting solvent extracts onto a series of silica coated glass rods. The rods are then placed sequentially into three solvent tanks, equating to aliphatics, aromatics and asphaltenes/polars (see Fig. 6.12).

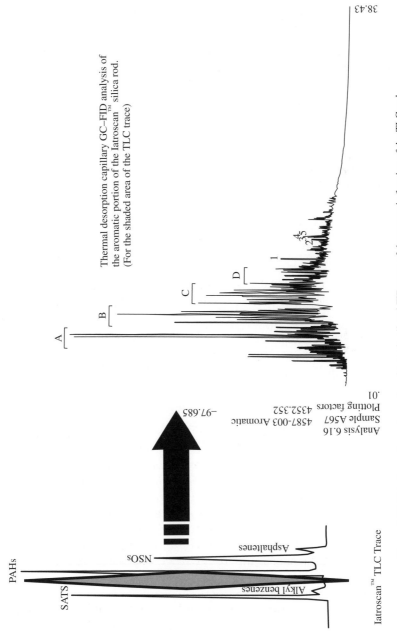

Figure 6.12 Thin layer chromatography of a crude oil and FID trace of the aromatic fraction of the TLC rod.

6.7.4.4   High pressure liquid chromatography (HPLC)

Solvent extracts are analysed by an HPLC system with a UV and/or fluorescence detector. The chromatogram can produce either a total PAH result, or can be fine-tuned to give speciated compounds. The method is specific and sensitive, but may suffer from co-elution, and the limit of detection for each PAH is ~1 mg/kg. For soil analysis, this is not the preferred technique as build up of matrix components on the HPLC column can result in small shifts in retention times leading to mis-identification of peaks.

6.7.4.5   Gas chromatography–flame ionisation detector (GC–FID)

Solvent extracts are passed through Florosil compounds to remove the aliphatic and resin fractions. The resulting eluate is analysed through a GC–FID system and the chromatogram integrated to give either speciated or total PAHs. The method is sensitive and specific, to a detection limit of 1 mg/kg, for each PAH.

6.7.4.6   Gas chromatography–mass spectrometry (GC–MS)

This is the most sensitive and specific method for PAHs, and involves analysing the solvent extract by a GC–MS system. The instrument is run in selected ion monitoring (SIM) mode, and therefore will only quantify the peaks of interest, ignoring any other compounds present. A less rigorous clean-up or no clean-up of the extract is required to avoid the build up of residues requiring time consuming column chops. However with *dirty* samples, this can also lead to build up of residues on the injector and detector. In addition many of the hydrocarbons present, often in very large quantities relative to the PAHs present, will be aliphatic and elute over the same range as the PAHs. This can degrade the accuracy of the quantitation of the PAHs. The method will analyse the sixteen or nineteen speciated PAHs to a level below 1 mg/kg. The number of target ions can be increased to cover isomers of the main two, three, four-ring compounds (i.e. methyl naphthalenes, dimethyl naphthalenes, etc.).

Abbreviations

BTEX	benzene, toluene, ethylbenzene and xylenes
DRO	diesel range organics
EDB	ethylene dibromide
FID	flame ionisation detector
GC	gas chromatography
GC–MS	gas chromatography/mass spectrometry
GRO	gasoline range organics
HPLC	high performance liquid chromatography
IR	infrared
MMT	methylcyclopentadienyl manganese tricarbonyl

MTBE	methyl tertiary-butyl ether
NSO	Nitrogen, sulfur or oxygen species
O&G	oil and grease
PAH	polycyclic aromatic hydrocarbons
PPM	part per million
SPE	solid phase extraction
SVOC	semi-volatile organic compound
TEL	tetraethyl lead
TLC	thin layer chromatography
TPH	total petroleum hydrocarbons
TPHCWG	total petroleum hydrocarbons criteria working group
TPHD	total petroleum hydrocarbons-diesel
TPHG	total petroleum hydrocarbons-gasoline
TPHV	total petroleum hydrocarbons volatiles
TRPH	total recoverable petroleum hydrocarbons
UCM	unresolved complex mixture
USEPA	United States environmental protection agency
UV	ultraviolet
VOC	volatile organic compound
VPH	volatile petroleum hydrocarbons

Glossary of terms

Additive	A substance added to petroleum mixtures (e.g. oxygenates) to impart new or to improve existing characteristics.
Aliphatic hydrocarbon	Hydrocarbons in which the carbon–hydrogen groupings are arranged in open chains (that might include branches). The term includes paraffins and olefins and provides a distinction from aromatics and naphthenes, which have at least some of their carbon atoms arranged in closed chains or rings.
Alkanes	Hydrocarbons that contain only single carbon–carbon bonds. The chemical name indicates the number of carbon atoms and ends with the suffix -ane.
Alkenes	Hydrocarbons that contain carbon–carbon double bonds. The chemical name indicates the number of carbon atoms and ends with the suffix -ene.
Alkyl groups	A group of carbon and hydrogen atoms that branch from the main carbon chain or ring in a hydrocarbon molecule. The simplest alkyl group, a methyl group, is a carbon atom attached to three hydrogen atoms.
Alkyne	Hydrocarbons that contain carbon–carbon triple bonds.

Analyte	The chemical for which a sample is tested, or analysed.
Antibody	A molecule having chemically reactive sites specific for certain other molecules.
Aromatic	A compound containing one or more benzene rings that also may contain sulfur, nitrogen and oxygen.
Asphaltene	Constituents of petroleum products with a high molecular mass. Chemically they are PAHs linked by aliphatic chains or rings and functional groups.
Biogenic	Bacterial or vegetation derived material.
Boiling point	A characteristic physical property of a liquid at which the vapour pressure is equal to that of the atmosphere and the liquid is converted to a gas.
BTEX	Benzene, toluene, ethylbenzene, and the three xylene isomers.
Bunker fuel	Heavy residual oil also called bunker C, bunker C fuel oil, or bunker oil.
Chromatogram	The resultant electrical output of sample components passing through a detection system following chromatographic separation. A chromatogram may also be called a trace.
Clean-up	A preparatory step following extraction of a sample media designed to remove components that may interfere with subsequent analytical measurements.
Cracking	A process whereby the relative proportion of lighter or more volatile components of an oil is increased by changing the chemical structure of the constituent hydrocarbons.
Crude oil	Naturally occurring mixture consisting essentially of many types of hydrocarbon, but also containing sulfur, nitrogen or oxygen derivatives. Crude oil may be a paraffinic, asphaltic or mixed based, depending on the presence of paraffin wax and bitumen in the residue after atmospheric distillation. Crude oil composition varies according to the geological strata of its origin.
Cut	The distillate obtained between two given temperatures during a distillation process.
Cycloalkane	A class of alkanes that are in the form of a ring.
Cycloparaffin	An example of a cycloalkane.
Diesel fuel	That portion of crude oil that distills out within the temperature range of approximately 200–370°C. A general term covering oils used as fuel in diesel and other compression ignition engines.
Distillate	A product obtained by condensing the vapours evolved when a liquid is boiled and collecting the condensation in a receiver that is separate from the boiling vessel.

Distillation range	A single pure substance has one definite boiling point at a given pressure. A mixture of substances, however, exhibits a range of temperatures over which boiling or distillation commences, proceeds and finishes. This range of temperatures, determined by means of standard apparatus, is termed the *distillation* or *boiling* range.
Eluate	The solutes, or analytes, moved through a chromatographic column (see elution).
Eluent	Solvent used to elute sample.
Elution	Process whereby a solute is moved through a chromatographic column by a solvent (liquid or gas), or eluent.
Extract	In solvent extraction, the portion of a sample preferentially dissolved by the solvent and recovered by physically separating the solvent.
Fingerprint analysis	A direct injection GC–FID analysis in which the detector output – the chromatogram – is compared to chromatograms of reference materials as an aid to product identification.
Flame ionisation detector (FID)	A detector for a gas chromatograph that measures anything that can burn.
Fuel oil	A general term applied to oil used for the production of power or heat. In a more restricted sense, it is applied to any petroleum product that is used as boiler fuel or in industrial furnaces. These oils are normally residues, but blends of distillates and residues are also used as fuel oil. The wider term, *liquid fuel* is sometimes used, but the term *fuel oil* is preferred.
Gas chromatography	An analytical technique, employing a gaseous mobile phase, that separates mixtures into their individual components.
Gas oil	A petroleum distillate with a viscosity and distillation range intermediate between those of kerosene and light lubricating oil.
Gasoline (petrol)	Refined petroleum distillate, normally boiling within the limits of 30–220°C, which, combined with certain additives, is used as fuel for spark-ignition engines. By extension, the term is also applied to other products that boil within this range.
Gravimetric	Gravimetric methods weigh a residue.
Grease	A semisolid or solid lubricant consisting of a stabilised mixture of mineral, fatty or synthetic oil with soaps, metal salts, or other thickeners.

Headspace	The vapour space above a sample into which volatile molecules evaporate. Certain methods sample this vapour.
Heating oil	Gas oil or fuel oil used for firing the boilers of central heating systems.
Hydraulic fluid	A fluid supplied for use in hydraulic systems. Low viscosity and low pour-point are desirable characteristics. Hydraulic fluids may be of petroleum or non-petroleum origin.
Hydrocarbons	Molecules that consist only of hydrogen and carbon atoms.
Immunoassay	Portable tests that take advantage of interaction between an antibody and specific analyte. Immunoassay tests are semi-quantitative and usually rely on colour changes of varying intensities to indicate relative concentrations.
Infrared spectroscopy	An analytical technique that quantifies the vibration (stretching and bending) that occurs when a molecule absorbs (heat) energy in the infrared region of the electromagnetic spectrum.
Jet fuel	Kerosene or gasoline/kerosene mixture for fuelling aircraft gas turbine engines.
Kerosene	A refined petroleum distillate intermediate in volatility between gasoline and gas oil. Its distillation range generally falls within the limits of 150 and 300°C. Its main uses are as a jet engine fuel, an illuminant, for heating purposes and as a fuel for certain types of internal combustion engines.
Light distillate	A term lacking precise meaning, but commonly applied to distillate, the final boiling point of which does not exceed 300°C.
Liquid chromatography	A chromatographic technique which employs a liquid mobile phase.
Liquid/liquid extraction	An extraction technique in which one liquid is shaken with or contacted by an extraction solvent to transfer molecules of interest into the solvent phase.
Mass spectrometer	An analytical technique that *fractures* organic compounds into characteristic *fragments* based on functional groups that have a specific mass-to-charge ratio.
Middle distillate	One of the distillates obtained between kerosene and lubricating oil fractions in the refining processes. These include light fuel oils and diesel fuels.

Mineral hydrocarbons	Petroleum hydrocarbons considered *mineral* because they come from the earth rather than from plants or animals.
Mobile phase	In chromatography, the phase (gaseous or liquid) responsible for moving an introduced sample through a porous medium to separate components of interest.
Naphtha	Straight-run gasoline fractions boiling below kerosene and frequently used as a feedstock for reforming processes. Also known as heavy benzene or heavy gasoline.
Naphthene	Petroleum industry term for a cycloparaffin (cyclo-alkane).
NSO	Organic compounds containing nitrogen, sulfur and/or oxygen.
Olefin	Synonymous with alkene.
Oxygenated gasolines	Gasolines with added ethers or alcohols, formulated according to the Federal Clean Air Act to reduce carbon monoxide emissions during winter months.
Polycyclic aromatic hydrocarbons (PAHs)	PAHs are a suite of compounds with two or more benzene rings. PAHs are found in many petroleum mixtures, and they are predominantly introduced to the environment through natural and anthropogenic combustion processes.
Paraffin (alkenes)	One of a series of saturated aliphatic hydrocarbons, the lowest numbers of which are methane, ethane and propane. The higher homologues are solid waxes.
Partitioning	In chromatography, the physical act of a solute having different affinities for the stationary and mobile phases. Partition ratios, K, are defined as the ratio of total analytical concentration of a solute in the stationary phase, C_S to its concentration in the mobile phase C_M.
Positive bias	A result that is incorrect and too high.
Purge and trap	A chromatographic sample introduction technique in volatile components which are *purged* from a liquid medium by bubbling gas through it. The components are then concentrated by *trapping* them on a short intermediate column, which is subsequently heated to drive the components on to the analytical column for separation.
Purge gas	Typically helium or nitrogen, used to remove analytes from the sample matrix in purge/trap extractions.

Retention time	The time it takes for an eluate to move through a chromatographic system and reach the detector. Retention times are reproducible and can therefore be compared to a standard for analyte identification.
Separatory funnel	Glassware shaped like a funnel with a stoppered rounded top and a valve at the tapered bottom, used for liquid/liquid separations.
Solvent	Fluids in which certain kinds of molecules dissolve. While they typically are liquids with low boiling points, they may include high-boiling liquids, supercritical fluids or gases.
Sonication	A physical technique employing ultrasound to intensely vibrate a sample media in extracting solvent and to maximise solvent/analyte interactions.
Soxhlet extraction	An extraction technique or solids in which the sample is repeatedly contacted with solvent over several hours, increasing the extraction efficiency.
Stationary phase	In chromatography, the porous solid or liquid phase through which an introduced sample passes. The different affinities the stationary phase has for a sample allow the components in the sample to be separated or resolved.
Supercritical fluid extraction	An extraction method where the extraction fluid, usually CO_2 is present at a pressure and temperature above its critical point.
Target analyte	Target analytes are compounds that are required analytes in USEPA analytical methods. BTEX and PAHs are examples of petroleum related compounds that are target analytes in USEPA methods.
Thin layer chromatography (TLC)	A chromatographic technique employing a porous medium of glass coated with a stationary phase. An extract is spotted near the bottom of the medium and placed in a chamber with solvent (mobile phase). The solvent moves up the medium and separates the components of the extract, based on affinities for the medium and solvent.
TPH E	Gas chromatographic test for TPH extractable organic compounds.

TPHD (DRO)	Gas chromatographic test for TPH diesel-range organics.
TPHG (GRO)	Gas chromatographic test for TPH gasoline-range organics.
TPHV	Gas chromatographic test for TPH volatile organic compounds.
Unresolved complex mixture (UCM)	The thousands of compounds that a gas chromatograph is unable to fully resolve into peaks. In the saturate fraction these are usually branched cycloalkanes.
Volatile compounds	'*Volatile*' is relative. It may mean (1) any compound which will purge, (2) any compound which will elute before the solvent peak (usually those $<C_6$), or (3) any compound which will evaporate during a solvent removal stage.
Wax	Waxes of petroleum origin consist primarily of normal paraffins. Waxes of plant origin consist of esters of unsaturated fatty acids.

Bibliography

ASTM Section 5 – Petroleum Products, Lubricants, and Fossil Fuels (2002) Volume 05.01 – Petroleum Products and Lubricants (I): D56–D2596. Volume 05.02 – Petroleum Products and Lubricants (II): D2597–D4927. Volume 05.03 – Petroleum Products and Lubricants (III): D4928–D5950. Volume 05.04 – Petroleum Products and Lubricants (IV): D5966–latest.

Barnabas, I.J., Dean, J.R., Fowlis, I.A. & Owen, S.P. (1995) Extraction of polycyclic aromatic hydrocarbons from highly contaminated soils using microwave energy. *Analyst*, London, **120**(7), 1897.

Bharati, S., Rostum, G.A. & Loberg, R. (1993) Calibration and standardisation of Iatroscan (TLC-FID) using standard derived from crude oils. *Advances in Organic Geochemistry*, **22**, 835–862.

Christensen, L.B. & Larsen, T.H. (1993) Method for determining the age of diesel oil spills in the soil, Groundwater Monitoring Report, Fall 1993, pp. 142–149.

Coleman, W.E., Munch, J.W., Streicher, R.P., Ringhand, H.P. & Kopler, F.C. (1984) The identification and measurement of components in gasoline, kerosene, and No 2 fuel oil that partition into the aqueous phase after mixing. *Archives of Environmental Contamination and Toxicology*, **13**, 171–178.

Dean, J.R., Barnabas, I.J. & Fowlis, I.A. (1995) Extraction of polycyclic aromatic hydrocarbons from highly contaminated soils: a comparison between soxhlet, microwave and supercritial fluid extraction techniques. Analytical proceedings – Royal Society of Chemistry, **32**(8), 305.

Drews, A.W. (ed.) (1992) ASTM Manual on hydrocarbon analysis 5th edition. ISBN 0803117671.

EPA 821/B-94-004b Method 1664: N-Hexane Extractable Material (HEM) and Silica Gel Treated N-Hexane Extractable Material (SGT-HEM) by Extraction and Gravimetry (Oil and Grease and Total Petroleum Hydrocarbons). Revised April 1995. (NTIS/PB95-239232).

EPA (1986) Method 8240, Gas Chromatography/Mass Spectrometry for Volatile Organics, Test Methods for Evaluating Solid Wastes, Volume 1b.

Heath, S., Koblis, K. & Sager, S.L. (1993) Review of chemical, physical and toxicological properties of components of total petroleum hydrocarbons. *Journal of Soil Contamination*, **2**(1), 1–25.

Institute of Petroleum (ed.) (2001) Standard Methods for the Analysis and Testing of Petroleum and Related Products and British Standard 2000 Parts, 2 Volume Set, Standard Methods for the Analysis & Testing of Petroleum & Related Products & British Standard 2001 ISBN: 0471491772. Volume 1 – Methods IP 1 to 339. Volume 2 – Methods IP 342 to 469, proposed methods, appendices. 2nd Edition.

Pollard, S.J. & Hrudey, S.E. (1992) Hydrocarbon wastes at petroleum- and creosote-contaminated sites: rapid characterisation of component classes by thin-layer chromatography with flame ionisation detection. ES&T, **26**(12), 2528–2534.

Rhodes, I.A.L., Hinojosa, E.M., Barker, D.A. & Poole, R.L. (1994) Pitfalls using conventional TPH methods for source identification: a case study; Proceedings of the Seventeenth Annual EPA Conference on Analysis of Pollutants in the Environment Norfolk, VA, 1994 (EPA Report 821-R-95-008).

Whittaker, M. & Pollard, S.J.T. (1994) Characterisation of refractory wastes at hydrocarbon-contaminated sites: 1. Rapid column fractionation and thin layer chromatography of reference oils. *Journal of Planar Chromatography*, **7**(5), 354.

Editors' note

Petroleum hydrocarbons are a mixture of many substances. The results of petroleum analysis depend as much on the analytical method as the petroleum. As long as everyone works to a common definition there is no problem. The Total Petroleum Hydrocarbon Criteria Working Group (TPHCWG) convened in 1993 to develop scientifically defensible information for establishing soil cleanup levels that are protective of human health at petroleum contaminated sites. The TPHCWG compiled their data and analytical efforts into five volumes that can be accessed at http://www.aehs.com/publications/catalog/contents/tph.htm:

- Volume 1. Analysis of Petroleum Hydrocarbons in Environmental Media.
- Volume 2. Composition of Petroleum Mixtures.
- Volume 3. Selection of Representative Total Petroleum Hydrocarbon (TPH) Fractions based on Fate and Transport Considerations[1].
- Volume 4. Development of Fraction-Specific Reference Doses (RfDs) and Reference Concentrations (RfCs) for Total Petroleum Hydrocarbons (TPH).
- Volume 5. Human Health Risk-Based Evaluation of Petroleum Contaminated Sites: Implementation of the Working Group Approach[2].

The above represent the most widely available generic approach to hydrocarbon assessment. Risks from carcinogenic components (BTEX & PAH) are assessed first as they usually drive the risks. The remaining petroleum hydrocarbons are assessed as a series of 13 fractions defined on the basis of 'Equivalent carbon' (EC) numbers rather than 'carbon numbers'. The EC number is related to the boiling point of individual compounds and retention time on a GC column. For example, the EC number of benzene is 6.5 because its boiling point and GC retention time are approximately halfway between those of *n*-hexane (6-carbon chain) and *n*-heptane (7-carbon chain). The TPHCWG chose the concept of EC numbers because these values are more logically related to compound mobility in the environment than carbon numbers.

---

[1]Volume 3 defines fractions of TPH that are expected to behave similarly in the environment. Analysis of environmental samples, fate and transport modeling, and risk assessment of petroleum contaminated sites is carried out in terms of these fractions.

[2]Volume 5 integrates the findings of Volumes 1 to 4 into a risk based framework for development of assessment criteria at TPH contaminated sites.

7   Volatile organic compounds

Sue Owen and Peter Whittle

7.1   Introduction

The substances covered in this chapter have a broad range of properties, from hydrophobic through to hydrophilic, and a wide range of molecular weights. They do, however, have a commonality in that they have a significant volatility at ambient or slightly elevated temperatures, and typically cover an approximate boiling range of −25°C to +200°C. Vapour pressures and solubility in water are wide-ranging and, together with the variability of a soil matrix, have implications for sample handling and the method of analysis. Many of the properties of volatile organic compounds (VOCs) and associated analytical problems are very similar to those of semi-volatile organic compounds, and therefore some of the detailed discussions in the semi-volatile organic compounds (SVOCs) chapter will not be repeated here, but cross-reference will be made as appropriate and differences detailed. We shall not cover the analysis of gases.

This chapter discusses the presence of VOCs in soil, the three common methods of analysis in use – methanol extraction/gas chromatography, purge and trap/gas chromatography and static head-space/gas chromatography – plus other techniques and problems. The substances determined typically include the full range of aliphatic and aromatic hydrocarbons, from about C_5/C_6 through to C_{11}/C_{12} and including naphthalene. A wide range of halogenated hydrocarbons from vinyl chloride to hexachlorobutadiene can be determined. Polar compounds, particularly the low molecular weight alcohols, aldehydes and ketones are more difficult to determine by the purge and trap or headspace techniques due to their solubility in water. Some may be determined by the methanol extraction method or in other cases by direct injection of an aqueous extract. However because they are readily biodegraded and of low toxicity, there is generally little interest in these substances, except in specific circumstances such as a spillage.

7.2   Presence of VOCs in soil

The presence of organic compounds in soil and the difficulties of extraction are extensively discussed in the chapter on SVOCs. The primary sources of VOCs in soil are from petroleum products and the industrial use of solvents including both halogenated and non-halogenated products. VOC contaminants in soil may be present as liquid films, adsorbed on soil aggregates or absorbed within soil pores. Higher boiling point substances may be present as particulates. Contamination by volatile substances is largely direct rather than from diffuse sources via the atmosphere or agriculture. Unlike water systems where haloforms may be formed extensively as by-products of disinfection, the production of VOCs as a by-product of soil treatment processes is unlikely to be encountered. A rare exception might be the use of formaldehyde as a bactericide to prevent the spread of animal diseases.

VOCs of petroleum origin are frequent contaminants of soil which is unsurprising, given the huge quantity of petroleum that is in daily use. Typical problems include leakage from storage tanks, particularly underground tanks at retail outlets, spillage and run-off from garage forecourts and transport operations, leakage and spillage from industrial sites and refineries, vehicle scrap and recovery operations, airports, and occasionally oil transport pipelines. The types of oil contaminants include all the distillate fractions from white spirit through to gas oil/diesel, residual fuel oils, lubricating oils and many special fractions used as industrial solvents or for the production of synthetic chemicals, for example alkyl benzenes. The nature of the oil itself may change within the ground depending on a number of factors, e.g. exposure to the atmosphere and oxygen availability, temperature and nature of the soil matrix. These can result in loss of volatiles, particularly with the spirit fractions, resulting in a higher proportion of less volatile components. Biodegradability may also occur when conditions are suitable, particularly affecting the light aromatic components that are more readily degraded than alkanes. The net effects of these changes and situations are not only to make identification of the type of oil more difficult, but also the analysis in itself. Apart from direct aqueous injection, all the analytical techniques used for VOCs involve the partition of the VOCs either into a solvent or the vapour phase. The presence of other less volatile components in varying amounts, in addition to soil matrix effects, affects the partition ratio, increasing the uncertainty in the measured value (see also Chapter 6).

Industrial solvents, including chlorinated compounds are less common soil pollutants. With the exception of dry-cleaning solvents they tend to be more restricted to industrial areas and waste sites. Unlike the aromatic solvents they are very difficult to biodegrade, but their partition is also affected by the nature of the soil and presence of other organic pollutants, particularly oils.

7.3 Sampling and sub-sampling

The very nature of VOCs precludes a reproducible sampling and sub-sampling process. Indeed the requirement for small sample and sub-sample vessels, either for analytical reasons or to reduce losses, generally dictates a selective sampling process. The variability of the soil matrix and the inability to dry and grind a sample in order to get a representative sub-sample for analysis mean that the uncertainty associated with VOC sampling and analysis is probably greater than for any other parameter. Sub-sampling is a particularly loss inducing process. Storage in containers, unless analysed in their entirety, introduces further losses on sub-sampling, by exposing samples to the atmosphere. The sensitivity of VOC analysis techniques requires only small samples, typically of the order of 1 g or so. This additionally means that sampling selectivity is introduced in taking a sample into the analytical stage. Care must be taken in the analysis of biologically active sample soils with high organic content such as sludge amended soils. Not only will some compounds biodegrade, reducing the measured concentration, but cross reactions within the sample may occur resulting in the formation of new substances. Typical of this process would be the interaction of hydrogen sulfide with certain ketones, to produce organo-sulfur compounds. Hence, VOC analysis should always be carried out as soon as possible after sampling. For this reason, and to minimise other losses, analysis for VOCs should always be undertaken with minimum delay between sampling and analysis. Samples should, wherever possible, be refrigerated from the time of sampling to the analytical process.

Contamination may occur during sampling, transportation and from contaminated laboratory air. Owing to the volatility of the determinands and the nature of the purge and trap procedure, considerable care needs to be taken to avoid losses by evaporation. Equipment or samples may be contaminated as a result of working in an atmosphere where solvents are in use. A number of precautions are detailed in this chapter to help minimise these problems.

Contamination may arise from use of solvents within the laboratory or from an adjacent laboratory with a shared ventilation system and can lead to airborne contamination of sample vials and other equipment. Particular care should be taken in the area where samples are handled and transferred and during the preparation of concentrated standard solutions. Many laboratories, particularly where large volumes of solvents are regularly used, find it necessary to have special room, often positively pressurised, for the preparation and analysis of samples for VOCs. Recently decorated rooms can also be a source of VOCs from surface coatings. Contamination can often be intermittent with the wind direction being the controlling factor.

Plastics can also present a problem from two points of view. It is not generally realised that many plastics not only absorb VOCs, but equilibrium is set up between the plastic and adjacent gas or liquid phase. Thus plastics in contact with standards may decrease or increase the concentration in the standard dependent on the level of contamination in the plastic. Blanks and samples can be similarly affected. Wherever possible therefore, the use of plastics is minimised and restricted to PTFE or similar material.

7.3.1   Artefacts

Despite the very best precautions, samples may still exhibit contamination and the presence of artefacts. Blanks should be reviewed regularly and each laboratory should familiarise themselves with the types of compounds and concentrations commonly detected in their own systems.

Artefacts may arise from sample and blank contamination, reactions and breakdown products within the chromatographic system. Examples of typical artefacts include acetal, silicone compounds, squalene, phthalate esters and other plasticisers and anti-oxidants.

7.4   Methods of analysis

7.4.1   Methanol extraction

Samples requiring analysis for VOCs should be taken from an untreated field-moist-soil sample wherever possible.

The test sample is extracted with methanol. This extraction is better done in the field with the sample being added into a pre-weighed screw-cap glass vessel, containing a known amount of methanol. This does, however, make selectivity in field sampling inevitable. However, losses of volatiles can be significant when samples are transported to a laboratory, even in sealed containers, without the addition of the extractant solvent. An aliquot of the methanol extract is then added into a headspace vial with a known amount of water and sealed. The temperature of the vials is stabilised in a thermostatic system to within the range 50–80°C to achieve specified equilibrium conditions. Gas chromatographic analysis of the volatile compounds in gaseous phase in equilibrium with the water in the vials is carried out using headspace injection with separation being achieved by the use of a capillary column with an immobile phase of low polarity. Volatile aliphatic and aromatic hydrocarbons may be detected with a flame ionisation detector (FID) or photo ionisation detector (PID). Volatile halogenated hydrocarbons are detected with an electron capture detector (ECD). Identification and quantification takes place by comparison of retention times and peak heights (or peak areas) towards internal standard added with the corresponding variables of an external standard solution. Volatile aromatic hydrocarbons and volatile halogenated hydrocarbons can be determined in one gas chromatographic run with the use of mass spectrometric detection. The efficiency of the procedure depends on the composition of the soil that is investigated, but improvements in partition can be achieved by use of a saline solution, rather than water, in the headspace vial. The quantitation procedure does not take into account incomplete extraction caused by structure and composition of the soil sample.

7.4.2 Head-space

Samples of (*as received*) soil or water associated with contaminated land are placed in glass vessels that are sealed. Sufficient headspace is left in the vials, which are then allowed to reach constant temperature equilibrium, where the concentration in the headspace is equal to the concentration in the matrix. An aliquot of this headspace is analysed directly by gas chromatography using a suitable detector or, preferably, GC–MS. Quantitation is carried out by comparison of the results for aqueous samples to which a known amount of the target analytes has been added. Limitations encountered using this technique include: extraction efficiencies from soils containing a large amount of clay or organic matter may be poor; equilibrium may not be reached when a soil sample is saturated with VOCs; high concentrations of analytes may lead to carry over in subsequent samples.

Analysis using headspace-gas is a very convenient way of cleaning up a sample before the actual GC analysis. It is far more acceptable for samples not able to be handled using a syringe, such as soil samples. Analytical procedures based on headspace GC are becoming increasingly popular with more and more trace determinations of volatile compounds needed due to the increasing number of ecological problems nowadays. It is preferred if standard GC procedures cause problems with the samples matrix in respect of solubility or thermal stability.

7.4.3 Purge and trap

An aliquot of sample is purged with an inert gas to strip out the volatile components that are subsequently trapped on an adsorbent column. The trap is then heated to desorb the volatile components which are swept by carrier gas on to a capillary GC column, separated utilising temperature programming and detected by the use of a mass spectrometer. The data are acquired in full scan mode so that the spectra may be matched against those of known standards. Quantification is carried out using selected characteristic m/z values for each determinand.

Limitations of the technique include: non suitability for non-purgeable substances and some polar compounds; high cost and maintenance; frequent cleaning of glassware compared to headspace; contamination of gas lines from highly contaminated samples; foaming from samples containing surfactants. However on the positive side, modern chromatography data handling systems have simplified the quantitation and identification process, enabling the determination of large suites of analytes.

7.4.4 Direct aqueous injection

This technique, although widely used, and indeed necessarily used for very polar solvents, is rarely used for the analysis of VOCs in soils due to the extensive contamination of injectors and columns by co-extracted materials. The analysis of soils for highly polar solvents is rarely requested, as these compounds are generally readily biodegradable.

7.5   Chromatography

Gas–liquid chromatography involves a sample being vapourised and injected on to a chromatographic column. The sample is transported through the column by the flow of inert, gaseous mobile phase. The column itself contains a liquid stationary phase which is adsorbed on to the surface of an inert solid. The carrier gas must be chemically inert. Commonly used gases include nitrogen, helium, hydrogen, argon and carbon dioxide. The choice of carrier gas is often dependent upon the type of detector that is used. The carrier gas system also contains a molecular sieve to remove water and other impurities.

There are numerous textbooks on and guides to chromatography, it is not intended to discuss the details of this technique in this chapter. The use of capillary columns, particularly in conjunction with headspace or purge and trap analysis, and modern bench-top mass spectrometers with large desk-top computing capacity, has produced a powerful technique for the analysis of VOCs. Typically, the separation, identification and quantification of up to about one hundred compounds in a single analysis are possible and routinely carried out.

7.5.1   Capillary columns

Fused silica is the most common inert and temperature stable material used in the manufacture of capillary columns. They are flexible, robust and easy to install. Many laboratories find difficulty in separating the ever increasing list of volatile compounds requiring analysis. Many phases suffer from low thermal stability, and analysis times are unacceptably long. The most common column used in the analysis of VOCs is the 624 column, available from several suppliers worldwide.

7.6   Detectors and quantitation

7.6.1   Mass spectrometric detection (MS)

Mass spectrometry is becoming the most common detector for the analysis of VOCs in soil samples, particularly as bench top instruments have increased in performance and decreased in price. Very often, the nature of the contaminants present is unknown and MS is used to identify individual compounds in the sample. Mass spectrometry may be used either in full scan mode or selected ion monitoring, depending on the nature of the compounds to be analysed and whether or not identification and quantitation are required. The use of selected ion monitoring provides a very sensitive technique; however, the concentrations of contaminants in soils can be extensive and difficulties may arise in that some substances can be above the calibration range of the instrument. Where the calibration range is exceeded, re-analysis would be required. When using methanol and aqueous extractions, these can be diluted with minimal losses, whereas with head-space and purge and trap techniques, further sub-sampling of the soil is required, possibly leading to further losses. Great care must be taken when using selected ion monitoring to ensure that compounds other than the target compounds are not missed.

7.6.2   Flame ionisation detector (FID)

The FID may be used when the contamination of a site/sample is known, e.g. BTEX (benzene, toluene, ethyl benzenes and styrene). Utilisation of the FID will reduce the cost of analysis without compromising the identification of target compounds. The FID generally has a poor response to chlorinated solvents. A major advantage of FID is the very wide dynamic range of the detector, enabling calibration over a wide range and hence reducing the amount of re-analysis required.

7.6.3   Electron capture detector (ECD)

The ECD is commonly used for halogenated compounds such as pesticides, polychlorinated biphenyls (PCBs) and chlorinated solvents where a lower detection limit than that obtainable by mass spectrometry is required. The ECD has a limited dynamic range and highly contaminated samples may compromise the calibration of the instrument. Depending on the nature of samples analysed, the detector will need regular maintenance and cleaning, as a detector becomes contaminated, the sensitivity increases but the dynamic range will decrease. The use of regular quality control samples will ensure that this situation is monitored. The ECD generally has a poor response to non-halogenated compounds, however, many compounds do exhibit significant responses e.g. polycyclic aromatic hydrocarbons (PAHs). Care must be taken to ensure that those compounds giving a poor response do not lead to false positive results in samples where they occur in high concentrations.

7.6.4   Electrolytic conductivity detector (ELCD)

The electrolytic conductivity detector (ELCD or Hall Detector) takes the chromatographic column effluent through a high temperature catalytic reactor converting halogens into inorganic acids that are detected by measuring changes in electrical conductivity in a flow through conductivity cell. The detector has been extensively used, particularly in the USA, for the detection of pesticides and other halogenated compounds, but has never found favour in the UK. The ELCD offers good sensitivity for halogen containing compounds and can also be set up for nitrogen and sulfur compounds (though not simultaneously), but early models were difficult to operate and maintain.

7.6.5   Photo-ionisation detector (PID)

The PID is well suited and quite widely used in environmental analysis, particularly where the main interest is in aromatic and other hydrocarbons. HNU Systems, Inc. developed the first PID in 1976 (Driscoll & Spaziani 1976). Molecules are ionised by absorbing UV photons of sufficient energy. The positively ionised molecules are accelerated away from a positively charged electrode towards a collector electrode at which the current is measured.

The PID is a non-destructive concentration sensitive detector that is 10–100 times more sensitive than the FID. The sensitivity increases with carbon number and aromatics > olefins > aliphatics. Certain solvents including water, methanol and pentane in particular, have a low sensitivity in a PID and therefore can be used as extractants. The dynamic range and good sensitivity towards certain compounds, reasonable selectivity, ease of use and robustness, have therefore made the PID a successful detector in some environmental situations, particularly those related to the petroleum industry.

7.6.6   Standards and quantitation

The preparation of standards for the analysis of VOCs does present difficulties for many laboratories. Firstly, the common use of many solvents in laboratories means that contamination and high blanks are real problems. The analytical techniques used are highly sensitive, standard solutions are usually prepared in methanol, and therefore absorption of atmospheric contaminants readily occurs. Secondly, the volatile nature of the components may make quantitative transfer susceptible to losses. For these reasons, and because of the large number of components involved, many laboratories prefer to buy in solutions of standard mixtures, which are usually prepared gravimetrically under clean conditions. However, laboratories need to check the accuracy of such solutions, usually by comparison to solutions from other suppliers and by means of proficiency schemes, and take care to minimise evaporative losses over the time-scale of use.

Care should be taken to monitor the manufacturer's uncertainty information in relation to the individual components. This can vary considerably, typically between ±0.5% and ±5% standard uncertainty. This information should influence the choice of manufacturer for calibration and quality control standards, with the lower uncertainty standards being used for calibration.

7.7   Screening techniques

7.7.1   Qualitative analysis

Due to the widespread use of solvents in the environment, an initial screening of samples for VOCs, either qualitatively or semi-qualitatively using mass spectrometric detection, has become a frequently used approach for identifying the type and extent of any pollution. For qualitative screening of samples the reconstructed total ion chromatogram (TIC) is generated and the peaks of interest selected. A check should be made that the peaks selected are not present in the blank at similar intensity.

A representative spectrum of each peak in turn is examined, after background subtraction if necessary, and submitted to a library search. An experienced GC–MS analyst must interpret the validity of the library matches. If the library match is considered to be unequivocal, the identification is reported. If isomer specificity is not certain, but elemental composition is considered unequivocal the fit is reported without specific isomer identifiers. If the compound type is unequivocal, but specific compound identification is not possible, then the compound class is reported. If no library matches are considered to be reasonable, then the component is designated as unknown. Individual laboratories should always establish their own criteria for reporting compounds as positively identified.

7.7.2   Semi-quantitative analysis

Qualitative analysis may be extended to provide *semi-quantitative* determination of concentrations. This may be carried out by addition of internal standards at known concentrations. Quantitation for each analyte detected is achieved by ratio of the total ion current for the analyte to that of the closest eluting internal standard. However, as the response of different compounds to mass selective detectors can vary considerably, this approach can lead to significant errors. A complementary approach that has found favour and can reduce the uncertainty involves the analysis of standards of tentatively identified compounds, in order to calculate response factors to the nearest internal standard. Standards are usually analysed several times over a period of time to calculate a mean response factor. This enables a semi-quantitative analysis of reasonable accuracy to be performed by reference to an in-house database of response factors. This eliminates the need for extensive calibrations for every batch and also reduces the errors from widely varying response factors.

7.8   Specific groupings

A frequently requested analytical suite is for BTEX – benzene, toluene, ethylbenzene and xylenes. This restricted suite of VOCs is widely used as an indicator of the presence of light petroleum pollution, i.e. mainly white spirit, petrol (gasoline) or kerosene pollution. The groupings that are commonly used when analysing samples for VOCs are given below. By-products of drinking water disinfection, trihalomethanes (THMs), are commonly found in potable water supplies worldwide. Chlorinated solvents from both dry cleaning and industrial use are common pollutants of groundwater. The analysis of THMs and chlorinated solvents is commonly carried out using GC–ECD.

References

Driscoll, J.N. & Spaziani, F.F. (1976) Development of a new photoionisation detector for gas chromatography. *Res. & Dev.*, **27**, 50.

Environment Agency (UK) (1988) Methods for the examination of waters and associated materials – The determination of volatile organic compounds in waters and complex matrices by purge and trap or by headspace techniques.

ISO 10381-1 (1994) Soil quality – Sampling – Part 1: Guidance on the design of sampling programmes.

ISO 10381-2 (1994) Soil quality – Sampling – Part 2: Guidance on the design of sampling techniques.

ISO 10381 (1997) Water quality – Determination of highly volatile halogenated hydrocarbons – Gas-chromatographic methods.

ISO 11465 (1993) Soil quality – Determination of dry matter and water content on a mass basis.

ISO 14507 (In preparation) Soil quality – Pre-treatment of samples for determination of organic contaminants.

ISO/DIS 15009 (In preparation) Soil quality – Gas chromatographic determination of the content of volatile aromatic hydrocarbons and halogenated hydrocarbons – Purge and trap method with thermal desorption.

Commonly used groupings for analysing VOC samples

Compounds which have been analysed by Purge and Trap/Headspace/GC–MS

benzene
bromobenzene
bromochloromethane
bromodichloromethane
bromoform
bromomethane
n-butyl benzene
sec-butylbenzene
tert-butylbenzene
carbon tetrachloride
chlorobenzene
chloroethane
chloromethane
2-chlorotoluene

4-chlorotoluene
1,2-dibromo-3-chloropropane
dibromochloromethane
1,2-dibromoethane
dibromomethane
1,2-dichlorobenzene
1,3-dichlorobenzene
1,4-dichlorobenzene
dichlorodifluoromethane
1,1-dichloroethane
1,2-dichloroethane
1,1-dichloroethene
cis-1,2-dichloroethane
trans-1,2-dichloroethane

1,2-dichloropropane
1,3-dichloropropane
2,2-dichloropropane
1,1-dichloropropene
ethyl benzene
hexachlorobutadiene
isopropyl benzene
p-isopropyl benzene
methylene chloride
naphthalene
n-propyl benzene
styrene
1,1,1,2-tetrachloroethane
tetrachloroethene

toluene
1,2,3-trichlorobenzene
1,2,4-trichlorobenzene
1,1,1-trichloroethane
1,1,2-trichloroethane
trichloroethene
trichlorofluoromethane
1,2,3-trichloropropane
1,2,4-trimethylbenzene
1,3,5-trimethylbenzene
vinyl chloride
o-xylene
m-xylene
p-xylene

Internal standards and surrogates

4-bromofluorobenzene
dibromofluoromethane
toluene-d_4
pentafluorobenzene

1,4-difluorobenzene
chlorobenzene-d_5
1,4-dichlorobenzene-d_4

Volatile internal standards with corresponding analytes assigned for quantitation

pentafluorobenzene
acetone
acrolein
acrylonitrile
bromochloromethane
bromomethane
2-butanone
carbon disulfide
chloroethane
chloroform
chloromethane
dichlorodifluoromethane
1,1-dichloroethane
1,1-dichloroethene
cis-1,2-dichloroethene
trans-1,2-dichloroethene
2,2-dichloropropane
iodomethane
methylene chloride
1,1,1-trichloroethane

trichlorofluoromethane
vinyl acetate
vinyl chloride

chlorobenzene-d_5
bromoform
chlorodibromomethane
chlorobenzene
1,3-dichloropropane
ethyl benzene
2-hexanone
styrene
1,1,1,2-tetrachloroethane
tetrachloroethene
xylene

1,4-difluorobenzene
benzene
bromodichloromethane
bromofluorobenzene (surrogate)

carbon tetrachloride
2-chloroethyl vinyl ether
1,2-dibromoethane
dibromomethane
1,2-dichloroethane
1,2-dichloroethane-d_4 (surrogate)
1,2-dichloropropane
1,1-dichloropropene
cis-1,3-dichloropropene
trans-1,3-dichloropropene
4-methyl-2-pentanone
toluene
toluene-d_8 (surrogate)
1,1,2-trichloroethane
trichloroethene

1,4-dichlorobenzene
bromobenzene
n-butyl benzene

sec-butyl benzene
tert-butyl benzene
2-chlorotoluene
4-chlorotoluene
1,2-dibromo-3-chloropropane
1,2-dichlorobenzene
1,3-dichlorobenzene
1,4-dichlorobenzene
hexachlorobutadiene
isopropyl benzene
p-isopropyl toluene
naphthalene
n-propylbenzene
1,1,2,2-tetrachloroethane
1,2,3-trichlorobenzene
1,2,4-trichlorobenzene
1,2,3-trichloropropane
1,2,4-trimethylbenzene
1,2,3-trimethylbenzene

BTEX

benzene
ethyl benzene
toluene

xylenes
MTBE (methyl tertiary butyl ether) is
   commonly analysed with this suite.

Tri-halomethanes (THMs) and halogenated solvents

bromoform
bromodichloromethane
carbon tetrachloride
chlorodibromomethane

chloroform
tetrachloroethene
trichloroethene
1,1,1-trichloroethene

Other compounds amenable to Purge and Trap analysis

acetone
acrolein
acrylonitrile
carbon disulfide
2-chloro-1,3-butadiene
2-chloroethylvinyl ether
p-dioxane
ethyl methacrylate
2-hexanone

iodomethane
methacrylonitrile
methyl ethyl ketone (MEK)
methyl methacrylate
4-methyl-2-pentanone
vinyl acetate
aliphatic hydrocarbons, pentane to
   decane

8 Non-halogenated organic compounds including semi-volatile organic compounds (SVOCs)

Joop Harmsen and Paul Frintrop

8.1 Introduction

The compounds described in this chapter cover a broad range of properties. They can be hydrophobic or hydrophilic, which may also be expressed by different distribution coefficients, K_{ow} and K_{oc}, between octanol and water or soil organic carbon and water. They also have different vapour pressures and solubilities in water. Due to the different properties, the interaction with the soil matrix also differs. Moreover, soil is not a single species; it contains mineral compounds of different sizes and properties as well as organic matter from various origins – all in different proportions. All the various compounds to be analysed and the diverse soil properties have implications for the pre-treatment of the sample, extraction or isolation of the analyte from the soil matrix, possible clean-up methods and the analytical procedure which can be used for quantification and qualification.

In this chapter, we first discuss the implications of soil and compound properties on sample pre-treatment and extraction procedures. Different detection methods can be used depending on compound features which are more or less specific. The effort necessary for complete identification is described. Examples of analyses of different groups of compounds are described in more detail.

8.2 Extraction

Extraction processes are quite complex and unfortunately there is not one universal procedure that is better than all others. Making a cup of coffee heavily relies on the quality of the extraction process. This analogy assists in understanding both the variety and the difficulties of extraction processes.

8.2.1 Presence of organic compounds in soil

Contaminated land can be the result of direct or diffuse pollution. This may strongly affect the way contaminants are present on the site and in the soil samples to be analysed. Direct pollution, as present in a lot of industrial sites, may result in the presence of the neat contamination (e.g. tar-like particles or liquids). In diffuse polluted areas, the contaminants are more evenly distributed and are present more like individual molecules adsorbed on or absorbed in soil. They can be introduced by air pollution, agricultural practice or long-term use of slightly polluted sludges. Contaminants, therefore can be present in soil in different forms. Rulkens (1992) gave a good overview of the different physici-chemical forms (Fig. 8.1). The contaminants can be present as particles, as liquid film, adsorbed on or absorbed in soil aggregates, present in large as well as in small pores. It is important to realise that the scale in soil, varies from tenths of microns in the case of absorption to tenths of millimetres in the case of particles and films. If organic compounds are bound to the soil they are prefer-entially bound to the soil organic matter with different strengths. If soil was a homogeneous material this binding could be described with chemical equilibrium, making use of distribution coefficients. From analytical point of view, the con-taminant can then be extracted with a solvent having enough attraction for the contaminant. Soil, however, is not a homogeneous material. On the molecular scale, the binding to the soil can be described as a normal chemical or physical interaction, but the interaction of soil aggregates with the surrounding liquid phase cannot. For instance, if the organic compound is absorbed in the soil organic matter it is more difficult for the extractant used to reach the contaminant than to dissolve the contaminant. Therefore, in soil the extractability is not only a func-tion of the preference of the compound for the solvent, but also of the access-ibility of the soil matrix. This results in compounds that are easily extractable (for instance present as separate droplets or films, adsorbed on the surface of soil aggregates) to very difficult extractable compounds that are adsorbed in very small soil pores or within organic matter. In Figure 8.2, this is illustrated as a distribution curve. Only a solvent that can reach all sites in the soil and that is strong enough to dissolve the compound will extract everything from the soil

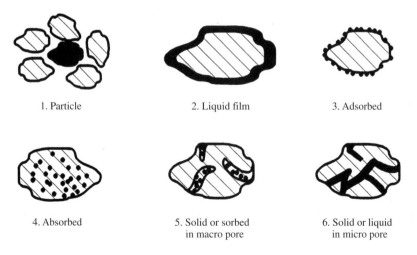

| 1. Particle | 2. Liquid film | 3. Adsorbed |
| 4. Absorbed | 5. Solid or sorbed in macro pore | 6. Solid or liquid in micro pore |

Figure 8.1 Appearance of contaminants in soil particles.

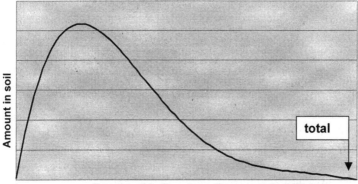

Figure 8.2 Distribution of a contaminant in soil over sites with different accessibility and binding strength.

matrix. If the attainability of the contaminant is low or if the contaminant is strongly bound to the soil matrix, less will be extracted (4 and 6 from Fig. 8.1 are present in the right part of Fig. 8.2). The amount of the contaminant under the right part of the curve will not be extracted using less strong extractants.

To isolate or extract a compound from the soil matrix the following steps are necessary:

1.   The extractant has to reach the compound (accessibility).
2.   The extractant has to release the compound (dissolving).
3.   The soil matrix has to be separated from the released compound.

In the past the accent was on step 2 *like dissolves like*. Good solvents were also considered as being good extractants. Steps 1 and 3 are not set by the properties of the organic compound, but by the soil matrix. The properties of organic contaminants differ from the properties of the soil. The way contaminants are present in soil has nothing to do with *like dissolves like*, but more with the fact that the contaminants have less attraction to their original environmental matrices (water and air).

For most organic contaminants, non-polar solvents like petroleum ether and toluene are very good solvents. These solvents, however, cannot reach every micro-pore in the soil. Selecting such a solvent may give good results if the contaminant is adsorbed on the surface of soil particles or present as solid particles or film. This is the case with fresh contaminants. These solvents also give good results with recovery experiments, as carried out during method development, because it is easy to extract freshly added contaminant. In aged polluted soil, however, the contaminant molecules do not stay on the surface, re-distribution takes place, resulting in diffusion to and adsorption on places (e.g. micro-pores) not reachable by this type of solvent. The accessibility can be increased by working with a more polar solvent that disintegrates the soil aggregates (like acetone) or by providing additional energy (e.g. higher temperature, higher pressure, longer extraction time, ultrasonic or microwave radiation).

The best extraction procedure is hardly predictable, due to the differences in types of soil and in the presence of the contaminants. Developing a new method should therefore compare the different potential methods. As mentioned before, it is not enough to measure only recoveries. The added compounds are easily accessible to the extractant and may even be extracted with a poor method. Certified reference materials (CRMs) are available for a variety of classes of compounds. Even if not all the analytes of interest are present in a CRM, the analyses may give a good indication of the capabilities of a tested extraction procedure. A further check of the extraction efficiency can be achieved by extracting the residual soil with other exhaustive procedures like Soxhlet. If the amount found is negligible compared to the amount obtained in the first extraction, it may be concluded that the first one is sufficient. Extraction methods that can be used are:

8.2.1.1 Soxhlet

Soxhlet has been developed for the determination of macro-compounds e.g. lipid content in food. It is a more or less continuous extraction procedure and it is regarded as an exhaustive extraction. The idea of an exhaustive extraction is very appealing for both macro and micro pollutants. Therefore *overnight* extraction i.e. extraction times of 14–16 hours are quite common with Soxhlet extraction. For micro-pollutants, such long extraction times may have some negative implications. This is caused by the non-specificity of any extraction procedure. Suppose that after n cycles, 70% of the available analytes and 10% of the more polar interferences have been extracted. The ratio between wanted and unwanted compounds is then 7. After 2n cycles, the recovery of the analytes has increased to 91% ($70+0.7\times30$) but at the same time the interferences have increased to 19% ($10+0.1\times90$). The ratio between wanted and unwanted has decreased to 4.8. It is clear that *infinite* extraction will end in an extract with quite a high content of unwanted substances. For the extraction of micro-pollutants, the power of the exhaustive extraction is at the same time its pitfall.

Soxhlet extraction does not apply any additional forces. The sample is soaked in the extraction liquid and with each cycle this liquid is supposed to drain completely from the sample. This procedure requires that the grain size of the sample has to be small in order to have a sufficient accessibility for the extraction liquid, but on the other hand the sample should not become too tightly packed because in that case only the outer surface will be extracted. A good check of the accessibility is the addition of some internal standard using a micro-litre syringe and injecting the internal standard through the wall of the extraction thimble at half the height of the sample bed into the centre of the sample. Complete recovery of the internal standard should not require more than 2–3 cycles.

In general, there are two procedures if non-polar micro-pollutants have to be extracted. The most common procedure is using a (chemically) dried sample, because water will prevent the extraction liquid from reaching the analytes. Magnesium sulfate monohydrate has proven to be a very effective drying agent in combination with Soxhlet extraction. Grinding the field-moist soil sample in a mortar with the same mass of magnesium sulfate monohydrate will result in a nice free-flowing powder with a good accessibility for solvents like petroleum ether or dichloromethane. The use of anhydrous magnesium sulfate is not recommended because drying with anhydrous magnesium sulfate generates much more heat compared with magnesium sulfate monohydrate. The drying capacity of the monohydrate is still about 75% of its own weight. Drying with the less expensive anhydrous sodium sulfate results in brittle lumps that require much more grinding.

For the more sandy soils with low water content, toluene may be used successfully due to the solubility of toluene in water and the azeotropic mixture of water in toluene. An example of this is the determination of PAH (see ISO 13877). Soxhlet extraction with azeotropic mixtures e.g. acetone n-hexane (60–40% m/m, bp 50°C) may be very useful if a more polar extraction solvent is required.

8.2.1.2 Shaking (agitation) procedure

Extraction by shaking is quite common in water analysis, even if the water sample contains appreciable amounts of suspended solids. Extraction by shaking has the advantage of both a proper mixing of the extraction liquid and the sample and the enhancement of accessibility by the applied mechanical forces. These benefits may also be obtained in the case of shaking soil samples with an appropriate extraction liquid. In the case of field-moist soil samples the extraction liquid has to be sufficiently miscible with water, because the mechanical forces itself are insufficient to provide accessibility for non-water miscible solvents like petroleum ether, pentane and hexane. Acetone and methanol are very well suitable to overcome the water barrier; they will break up the soil aggregates. The polarity of acetone and methanol is rather high, making them less suitable as solvents for non-polar compounds. Adding a less polar solvent that is miscible with acetone or methanol will overcome this. This means that in fact a two stage extraction is performed, first a rather short shaking with acetone or methanol only and after that a longer period of shaking with the mixture of acetone or methanol and petroleum ether, pentane or hexane. To be successful there are some boundary conditions; the ratio between acetone and the moisture in the sample shall be at least 9:1 in order to maintain the miscibility of the acetone–water phase with the non-polar hydrocarbon solvent.

Under normal circumstances the acetone will be removed from the extract by shaking the total extract with a large surplus of water, at least four times the volume of the acetone should be applied. Therefore, the amount of hydrocarbon solvent should not be too low. In general the volume of the hydrocarbon solvent should be 0.5–1 times the volume of the acetone, which means a ratio between hydrocarbon solvent and water acetone phase of 1/(5–10). It is this step that governs the final distribution of the extracted compound. The removal of the acetone will also imply a removal of other more polar compounds ($K_{ow} <$ ca 3). For the extraction of non-polar compounds, the removal of the acetone and other polar compounds may be regarded as a first clean-up of the extract.

Examples of the shaking procedure can be found in several ISO standards for the determination of organic micro-pollutants in soil, such as PAH, OCB and PCB, mineral oil and herbicides.

8.2.1.3 Pressurised fluid extraction (PFE)

Pressurised fluid extraction also known as accelerated solvent extraction (ASE) uses higher temperature and pressure. Going back to coffee it is the difference between espresso and filter coffee. By using a higher pressure and temperature the extraction becomes more effective and takes place in a shorter time. There is, however, a difference in extracted compounds. Comparing the efficiency with other methods, this can be positive, but also negative. As long as accessibility for the extractant is high enough, ASE will give high recoveries. ASE is a promising method for extraction of organic contaminants in soil. In some cases, ASE resulted in higher contents than certified values obtained by Soxhlet extraction. Therefore, the certified values of soil-CRMs should always be related to the extraction procedure used.

The extra input of energy by increasing the temperature and pressure makes it possible to reduce extraction times. However, if barriers for the solvent exist like presence of water or large soil aggregates, ASE may also fail. The first barrier can be overcome by using a water miscible solvent or adding methanol to a non-polar solvent or chemical drying of the sample. The use of (pure) acetone in ASE is not recommended because polymer formation may take place at the applied temperature and pressure, resulting in artefacts in the gas chromatogram. Large soil aggregates may limit the possibilities of ASE, because mechanical shaking in combination with a proper solvent is necessary to break them up.

The USEPA accepts ASE as an alternative for Soxhlet. EPA Method 3545A (1998) is a procedure for extracting water insoluble or slightly water soluble organic compounds from soils, clays, sediments, sludges and waste solids. The method uses elevated temperature (100–180°C) and pressure (1500–2000 psig) to achieve analyte recoveries equivalent to those from Soxhlet extraction, using less solvent and taking significantly less time than the Soxhlet procedure. This method is applicable to the extraction of semivolatile organic compounds, orga-nophosphorus pesticides, organochlorine pesticides, chlorinated herbicides, PCBs and PCDDs/PCDFs. The method is applicable to solid samples only, and is most effective on dry materials with small particle sizes. If possible, soil/sediment samples should be air-dried and ground to a fine powder prior to extraction. Alternatively, if the loss of analytes during drying is a concern, soil/sediment samples may be mixed with anhydrous sodium sulfate or pelletised diatomaceous earth. Drying and grinding samples containing PCDDs/PCDFs are not recommended, owing to safety concerns.

8.2.1.4   Super critical fluid extraction (SFE)

Super critical fluid extraction enlarges the range of solvents. Using super critical carbon dioxide has the advantage that the final extract is automatically concentrated to dryness. The polarity of supercritical carbon dioxide is rather low; therefore, small amounts of methanol are added as *modifier* to increase the polarity of the extractant. The solvents also have a large penetrating power because of their low viscosity. The invention of ASE has reduced the interest in SFE due to the more versatile applicability and robustness of ASE.

8.2.2   Pre-treatment in relation to extraction

Soils contain water, which has influence on the extractability of organic contaminants. The water covers the soil particles and is present in the micropores, which decreases the accessibility for the extractant. Especially when the solubility in water is negligible (non-polar components), the extraction efficiency will be low. Removal of water may also be prescribed if the homogeneity of the sample is low and crushing is necessary. The water has to be removed during pre-treatment (see Chapter 3). For organic contaminants this means chemical drying as described in ISO 14507 (2002). Another possibility is the use of water miscible solvents as the first extractant (e.g. acetone or methanol). This may be used when crushing is not necessary. Volatilisation or biodegradation during pre-treatment may occur or if the extraction procedure needs the presence of a certain amount of water. It has to be realised that if the water content is very high (e.g. peat, sludges and sediments), the extractant becomes diluted and less effective. Higher amounts of extractant will be necessary to be effective.

Organic compounds having acidic or basic properties often need adjustment of the pH, such as described for phenols (see Section 8.6.5). It may also be possible that some organic contaminants are more easy accessible and soluble in the presence of water as described for herbicides further in this chapter. In these cases, removal of the water is not permitted and the water present is part of the extraction procedure.

Component properties like the solubility and K_{ow} (distribution coefficient between octanol and water) can be used in the development of extraction procedures. Table 8.1 gives a overview of some components described in this book.

Table 8.1 Physical properties of organic contaminants (in unbuffered water unless marked otherwise)

Compound	Solubility in water (mg/l)	Log K_{ow}	Compound	Solubility in water (mg/l)	Log K_{ow}
Herbicides			**PAHs**		
Desisopropylatrazine	670	1.15	Naphthalene	31	3.37
Metamitron	1700	0.83	Acenaphtylene	3.9	4.07
Desethylatrazine	3200	1.51	Acenaphthene	3.8	4.33
Hexazinon	33 000	1.2 (pH 7)	Fluorene	1.9	4.18
Metoxuron	678	1.6	Phenanthrene	1.1	4.57
Bromacil	807 (pH 5)	1.88 (pH 5)	Antracene	0.045	4.54
Simazin	6.2 (pH 7)	2.1	Fluoranthene	0.26	5.22
Monuron	230	1.94	Pyrene	0.132	5.18
Cyanazin	171	2.1	Benz[a]anthracene	0.011	5.91
Methabenzthiazuron	59	2.64	Chrysene	0.002	5.75
Chlortoluron	70	2.41	Benzo[b]fluoranthene	0.001	6.57
Atrazine	33 (pH 7)	2.5	Benzo[k]fluoranthene	0.0005	6.84
Isoproturon	65	2.5	Benzo[a]pyrene	0.004	6.04
Diuron	36.4	2.85	Dibenz[a,h]anthracene	0.0006	6.75
Metobromuron	330	2.41	Benzo[g,h,i]perylene	0.003	7.23
Metazachlor	430	2.13 (pH 7)	Indeno[1,2,3-cd]pyrene	0.06	7.66
Terbutylazine	8.5	3.21	**PCBs**		
Propazin	8.6	2.93	PCB 28	0.27	5.62
Dichlobenil	14.6	2.7	PCB 31	0.14	5.69
Chloroxuron	3.7	3.70	PCB 44	0.1	5.81
Propyzamid	15	3.1	PCB 52	0.015	6.09
Terbutryn	22	3.65	PCB 101	0.015	6.80
Ethofumesat	50	2.7 (pH 7)	PCB 118	0.013	7.12
Metolachlor	488	2.9	PCB 138	0.0015	7.44
Alachlor	240	3.52	PCB 149	0.0042	7.28
Pendimethalin	0.28	5.18	PCB 153	0.00095	7.75
Organotin			PCB 170	0.0035	8.27
Triphenyltin	1.6	2.9–3.6 (pH dependent)	PCB 180	0.0038	8.27
Tributyltinchloride		3.4–4.2	PCB 194	0.00027	8.68
Fentinhydroxide	0.4	3.53	**Phenols**		
Tri-cyclo-hexyltin	0.02		Phenol	81 800	1.46
Tri-*n*-hexyltin	0.04		Methylphenol	9066	1.95
Dibutyltin		1.49	Ethylphenol	5340	2.47
			Trimethylphenol		3.15
Phthalates			Pentylphenol		3.95
Dimethylphthalate	4000	1.60	Octylphenol		5.50
Diethylphthalate	1080	2.42	Nonylphenol	6.35	5.99
Dipropylphthalate	108	3.27	**Chlorophenols**		
Dibutylphthalate	11	4.50	Monochlorophenol	28 500	2.16
Dipentylphthalate	100	5.62	Dichlorophenol	614	2.80
Dihexylphthalate	0.24	6.82	Trichlorophenol	121	3.45
Di(2-ethylhexyl)phthalate (dioctylphthalate)	0.27	7.60	Tetrachlorophenol	18	4.09
			Pentachlorophenol	14	5.12

These constants are available in Handbooks like the Pesticide Manual (Tomlin 2000) or can be found on the internet for instance on the Physprop database (http://esc.syrres.com/interkow/physdemo.htm). A good chemical entrance is www.chemfinder.com. If there are more than one isomer, figures for the most common isomers are given.

8.3   Screening or quantitative determination

In analytical environmental chemistry, the first question is whether an organic contaminant is present or not. Mostly this question is followed by a second question *how much*. The first question can be answered by using a screening method. Screening methods are used for a broad range of compounds. It is highly unlikely that the extraction procedure used in these types of methods will be very effective for all compounds. In a screening method, extraction procedures may be less efficient for part of the target compounds listed. This is illustrated in Table 8.2. Good and poor recoveries are found with a broad extraction procedure. Recoveries of relative polar ($K_{ow} < 2.5$) and non-polar compounds ($K_{ow} > 6$) are poor. Between these values of K_{ow}, recoveries are sufficient for a screening procedure. Analysis of specific compounds has to be carried out with extraction methods having high recoveries (in general >80%). With a broad extraction method that has to be used for screening purposes, low recoveries for some compounds have to be accepted.

Increase in number of extracted compounds = decrease of accuracy

In a screening method only two answers are accepted '*yes*' the compound is present or '*no*' the compound is not present. For the analytical scientist there are further issues which are caused by the uncertainty of a method. The result of an analytical method can be that a compound is present above a certain concentration with a certainty of 95%. The analyst also has to take into account

Table 8.2 Recoveries of some organic compounds in a clay and peat soil using a screening procedure. Extraction after wetting with ammonium chloride solution and extraction at pH 1 and pH 10 using acetone followed by petroleum ether/dichloromethane. Extracts are dried with anhydrous sodium sulfate, concentrated and injected in a GC–MS. Herbicides and chlorophenols are methylated using trimethylsulfoniumhydroxide (Traag *et al.* 1997)

Compound	K_{ow}	Recovery (%)	
		Clay	Peat
Phenol	1.46	50	0
Ortho-cresol	1.95	1	0
Atrazine	2.5 (pH = 7)	53	62
Mecoprop	3.13	82	44
Linuron	3.2	44	24
1-Chloronaphthalene	ca. 3.3	85	85
Naphthalene	3.37	100	121
2,4,5-Trichlorophenol	3.45	109	35
Lindane	3.72	86	99
Pentachlorophenol	5.12	125	46
Benzo(a)pyrene	6.04	7	52
p',p'-DDT	6.91	32	20

the low recovery of certain compounds. Are false negatives accepted if the compound is present, but reported as absent, or are false positives accepted if the compound is absent, but is reported as present. This is difficult to explain to regulators. The safest environmental option is the acceptance of some false positives. In this case one has to take into account the recovery. For example, if compounds with recoveries higher than 20% are accepted to be some of the target compounds. Thus absence can only be reported if the measured value $+2\times$ standard deviation is smaller than the target value/5. If the measured value $-2\times$ standard deviation is larger than the target value, the compound is really present. The area in between is a grey area, but the component is being reported as present. This area can be made smaller by using a more specific method with a higher recovery, a smaller standard deviation and measuring more samples and using the 95% confidence value. The most reliable results will be obtained by using a method specified for the compound tested as outlined in Chapter 2.

8.3.1 Group parameters

Another way of screening is the use of group parameters. Most often used is the extractable organic halogens (EOX) test. In this case, a validated extraction procedure is used and the halogens are quantified after combustion of the organic extract by a coulometric measurement of the halogens. Taking into account the different physical properties of organic halogens as presented in Table 8.1, the extraction procedure has to be prescribed exactly. Another extraction procedure will give different results, which is not acceptable in regulations (i.e. the method is empirical and defined by the procedure). The role of the EOX-measurement is a first step and may be followed by specific analyses (see also Section 8.6.1).

In principle, mineral oil (see also Chapter 6) is also a group parameter. All compounds that are extracted, are not removed during clean-up, pass the chromatographic column and are detected by the universal detector are said to be mineral oil. The influence on changing parameters of the method will not be as significant as with the EOX-determination. All mineral oil compounds are very non-polar. Hence care has to be taken.

Some other classes of compounds are also considered as suitable for being reported as group parameters. The final result is reported as a sum of the individual compounds. As long as the properties of the individual components and their interaction with the soil are comparable, they can be and are measured in one single method. Examples are PAHs (Chapter 6) and PCBs (Chapter 7). However, if there is a broad range in properties and the only similarity is in the name of the compound or group, it will be difficult to have a method with high recoveries for all of them. Phthalates and organotin compounds are examples of this and are discussed in this chapter.

8.4   The bioavailable fraction

In the scientific world, it is now common knowledge that measured total concentrations are insufficient to explain the environmental effects of contaminants. Part of the contaminants are so strongly bound to the soil matrix that they do not have any significant environmental effects. Bioavailability is a concept for which no uniform definition exists. Users with different backgrounds have provided different definitions as to what they consider to be bioavailability. From a chemical point of view, it is the amount per mass or volume unit that can desorb from soil or (suspended) sediment in a certain period of time under specific conditions. It can also be defined as the amount reversibly adsorbed to the soil, again under realistic boundary conditions. Ecologists define bioavailability as the amount (or concentration) to which organisms are actually exposed and that thus can be taken up, possibly causing an effect. It is, therefore, important to distinguish between bioavailability for human beings, plants and micro-organisms. Availability may also lead to bio-degradation of the contaminant. When taking into account also the *broader* environment, which includes groundwater, focus tends to be on the amount present or in equilibrium with the water phase of the soil which is roughly equivalent to the amount that can leach to lower soil horizons and ultimately the groundwater. There are changeovers in the soil, to organisms and within organisms to the target organ. In all approaches, the following sequence is part of the concept:

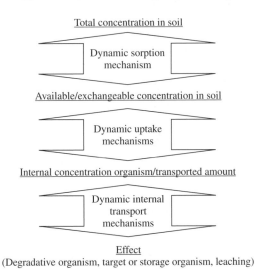

Total concentration in soil

Dynamic sorption mechanism

Available/exchangeable concentration in soil

Dynamic uptake mechanisms

Internal concentration organism/transported amount

Dynamic internal transport mechanisms

Effect
(Degradative organism, target or storage organism, leaching)

For various purposes, the dissolved concentration (or even the free ion activity) in the pore water can be used to explain effects. In those cases *available fraction* and pore water concentration are equivalent. This is, however, only one of the definitions of bioavailability. Based on the pore water concentration (*actual availability*) it is possible to explain *targets* such as the actual leaching, plant uptake and toxicological effects of pesticides just after application. It does not explain biodegradation, or effects on soil consuming organisms such as worms or uptake by little children. For these purposes it is necessary to include the *potential* bioavailability. If the actual availability is defined as the actual concentration or activity in the pore water or the equivalent of that obtained in a weak extraction, the potential availability is the amount of the contaminant that can become available within a certain time and depends on chemical conditions applied. The potentially available fraction always exceeds the actual determined available fraction but it cannot exceed the total concentration. For establishing the potential bioavailability, it is important to define the total system including the organisms and environmental conditions that influence the fate of a contaminant in the soil or the activity of organisms. Soil is part of such a system. For the organisms living in soil, a bio-influenced zone (Alexander *et al.* 2003) can be defined. This zone contains the pore water and depending on the organism, parts of the soil matrix itself (e.g. worms).

There are already several methods to measure the availability of contaminants. Measured values of various indicators of the chemical availability results in a close correlation with effects. In this chapter, we do not want to give an overview of all these methods. In a more general way Alexander *et al.* (2003) distinguished between the following chemical measures of bioavailability:

1. Water-based extractions

    - concentration in pore water.
    - extraction from the water (solid phase extraction).

2. Solvent-based extractions

    - mild solvent.
    - short extraction or extraction with low energy input.
    - mild acid or complexing agent (heavy metals).

Organic contaminants are almost completely adsorbed to the organic matter fraction. As long as the adsorption is reversible the partition between water and soil can be described using the K_{OC} value. For several contaminants this function is known or can be derived from K_{OW} or the water solubility. This reversible partition, however, only occurs with freshly added contaminant and is true for freshly added pesticides. In this (rather special) case, effects can be explained from the total concentration. With the more soluble pesticides (not including the old DDT-like types) it is also possible to use the pore water concentration. After ageing occurs, which results in a significantly reduced availability, it becomes impossible to explain the effects from the total concentration. For relatively soluble contaminants, the pore water concentration can still be used to explain internal concentrations and (possibly) effects.

For less soluble contaminants like PAHs and PCBs, it is not possible to measure pore water concentrations. Chemical availability has to be measured with a method that extracts a certain part from soil. Methods are available, such as a mild extraction for a short period of time. Relation with effects have been empirically established (certain strength and certain time). Methods are easy and cheap, but valid only after establishing the empirical relationship. Other methods are based on a strong adsorbent and stimulate the diffusion from the solid phase into the water phase and on to the adsorbent. This process is easier to understand (based on equilibrium between soil and water), but is also based on empirical relationships (Reid *et al.* 2000).

Although methods are available, they are still in an experimental phase. The bioavailable fraction is currently not part of regulations. It is however to be expected that there will be an increasing interest in bioavailable fraction in the coming years. For analytical laboratories this means that special empirical extraction procedures will be introduced that may be combined with existing measurements.

Comparing measurement of the bioavailable fraction with the total concentration, the method for the bioavailable fraction will measure mainly those contaminants presented in Figure 8.1 as being in liquid films, adsorbed on the surface and present in macro-pores. In Figure 8.2, it represented as most of the left part under the curve. This part cannot exactly be defined, because it depends on the organism involved. It will be relatively small for micro-organisms living in the water phase and larger for soil consuming organisms such as worms and also for small children.

8.5   Detection, identification and quantification

8.5.1   Detection

Most organic compounds are measured using chromatography, both liquid and gas chromatography. The first step is obtaining a good separation between the target compounds, but also a separation between target and non-target compounds. It may be necessary to do a clean-up on small columns or using multidimensional chromatography (LC/LC or LC/GC). The latter is not treated in this chapter but will become available for routine analyses in the future. After the separation, the compounds have to be detected. This can be done with a universal detection such as FID and UV, but also with more specific detectors such as ECD, NPD and fluorescence. A mass spectrometer (MS) is both universal and selective. As long as the separation power of the chromatographic system is high enough to separate the target compound from the background and the target compound is present in a relative high concentration, good results can be reached with a less specific detection. With complex samples a better selectivity can be obtained by:

- Improving the clean-up.
- Increasing the separation power of the chromatographic system.
- Using a more specific detector.

Laboratories are more and more focussed on the latter, using a more specific detector, the MS. This is a good development, however, this should *not* replace the clean-up and chromatographic system. Negligence of clean-up, with the argument that the detector is specific enough, will lead to contamination of the detector and reduction in separation power of the chromatographic system by adsorption of interfering compounds in the injection system and the column. The gain of the more specific analysis will be mitigated by the lack of or reduced cleanup of the obtained sample extract.

8.5.2 Identification

Use of an MS does not automatically lead to a reliable identification of the compound to be analysed. A certain number of conditions have to be fulfilled. Again a contradiction is present. The highest degree in identification is achieved if the complete mass spectrum is identical to the complete mass spectrum of the standard. A lot of MS systems however are most sensitive if only one mass is measured, but one mass is not very selective.

High selectivity = low sensitivity

Are there a number of limited conditions to have a proper identification? The procedure to qualify a component consists of two steps (ISO/DIS 22892):

Step 1: Chromatographic result
The relative retention time should fulfil the criteria. Only if step 1 is positive can step 2 be made. For GC-analyses these criteria can be very strict. For instance, the relative or absolute retention time measured in the sample should not differ by more than $\pm0.2\%$ (from the relative retention time in the last measured external standard solution). In HPLC analyses, a higher deviation has to be accepted.

Step 2: Gathering identification points
For qualification the principle of identification points is used (European Commission 2002). Identification points can be obtained from MS-data, but also from other sources of information. The following classification can be obtained. The flow scheme is presented in Figure 8.3:

- Identification: The target compound is present in the analysed extract. At least *three* identification points are obtained;
- Indication: The target compound may be present. *One* or *two* identification points are obtained;
- Absence (below detection limit): No identification points are obtained using MS.

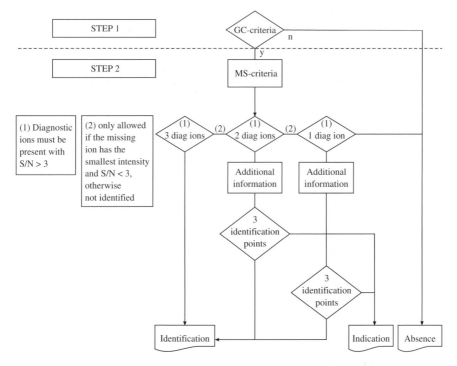

Figure 8.3 Flow scheme for the identification of a target compound using GC–MS.

Firstly the MS results are evaluated by comparing the relative intensities of the selected ion peaks. For every ion peak an identification point is obtained, when its relative intensity (compared to the base peak in %) measured in the sample does not deviate more than $\pm((0.1 \times I_{std}) + 10)\%$ from the relative intensity of the diagnostic ion in the external standard solution. Three identification points give a positive identification. If less than three ion peaks are available (due to sensitivity (S/N < 3) or absence of fragments (PAH)), additional identification points should be gathered using additional analytical evidence. The possibilities are given in Table 8.3.

Table 8.3 gives the analytical possibilities but it is also possible to use additional evidence. Interpretation of environmental data is always a combination of data analyses, knowledge about the origin of the sample, knowledge on the behaviour of contaminants and processes that occur or may occur. This is also true for the interpretation of GC–MS analysis. As stated a component is identified if three identification points are obtained. If only one or two diagnostic ions are present additional identification points are necessary. An extra identification point is obtained if one of the following criteria is fulfilled:

- The component is identified in earlier samples from the same site (for instance, if the sample under investigation has a low concentration and one or two diagnostic ions have $S/N < 3$ following the biodegradation).
- Other samples from the same site give positive identification.
- No other ions are visible in full scan mode and this is in agreement with the mass spectrum of the pure component (for instance PAH).
- Identification is in agreement with the pattern normally present or present on that site (for instance PCB or PAH).
- From historical investigation it was shown that presence of the component was expected.
- For volatile compounds the specificity of the mass fragments in combination with their retention time will generally be sufficient. Their volatility corresponds to a low molecular weight, limiting the number of possible false positive results: there are not many low molecular weight compounds with the same retention time on a GC-column and similar mass spectra.

8.5.3 Quantification

Measuring organic compounds, quantification can make use of internal and external standards. Different examples are described in this book. There is no general advice on the various organic compounds in soil that can be quantified. Due to the complex composition of soil, which influences the extractability, it is not customary to correct measured values for the incomplete extraction. There can be a high extraction efficiency for one soil and a lower extraction

Table 8.3 Examples of number of identification points (n = integer). Derived from European Commission (2002)

Source	Identification points (n)	Remark
Diagnostic ion	1	Every ion $S/N > 3$
Absence of any other ions in full scan	1	Diagnostic ions in full scan $S/N > 3$
Column with different polarity	1	GC-criterion (extra retention time value)
Isotope dilution	1	
Expectation, plusibility, earlier investigations		
Component spike/standard addition	1	
Chromatographic pattern	1	i.e. PCB, PAH, Dioxins
Other analytical techniques	1	Every other selective detector or technique
Other MS-techniques		
GC–MS (EI and CI; positive/negative)	4	2 (EI) + 2 (CI)
GC–MS–MS	4	1 precursor and 2 daughters
HR–MS (high resolution MS)	2	Every ion $S/N > 3$

efficiency for another similar soil. A quantitative soil analysis should be effective enough to extract the target compounds from most soils with an efficiency of more than 80%. In this case the measured concentration may be reported as the correct value.

Is this always true? Whilst everyone is using the same standardised method, one will never know. Also there should not be continual discussions on the best *in-house* method. We have to realise that there will always be a soil for which a well validated method will fail. Is this a real problem?

- No – If the measurement is carried out in relation to legislation. Because of the equality before the law, it is more important that the analyses are reproducible so that different laboratories will obtain the same results. Because the ultimate method does not exist, it has to be accepted that the standard method in use sometimes will give a too low concentration. A good method development and validation will diminish the chance on these *wrong* analyses but cannot prevent them all. We have to accept the failure of the method in a limited number of cases.
- Yes – If we want to know the effect of a contaminant in a specific soil. In that case we want to know the real concentration and cannot take the risk that our standard procedure is failing in our specific sample. For these types of investigations it will always be necessary to re-validate the method used with the relevant soil.

Some people will also say that there is a problem with analyses for legislation, but then they will have to prescribe that a specified standard method should be tested with every sample. This will make the cost of environmental analysis prohibitive.

8.6   Examples

8.6.1   Group parameters for organo-halogens

Organo-halogen group parameters like purgeable (POX), extractable (EOX) and activated-carbon adsorbable (AOX) organo-halogens are good indicators for man-made pollution. However for each group the interpretation of results in view of environmental effects is difficult, due to the large differences in physico-chemical and toxicological behaviour of the individual compounds within each group. The broader the range of compounds, the larger these differences are. The effect of, for example, $10\,\mu g/l$ as Cl of some humic acid molecule with some hydrogen atoms replaced by chlorine cannot be compared with $10\,\mu g/l$ as Cl of 2,3,7,8-TCDD. Quality standards for organo-halogen group parameters are therefore hardly found in legislation.

These group parameters have been extensively used in water analysis. AOX was developed in Germany especially for drinking water analysis and was transposed to the analysis of surface water and wastewater. In Germany, there exists a levy for AOX in wastewater. In the Netherlands, EOX was developed as an indicator for bio-accumulatable organo-halogen compounds. The use of, for example, n-hexane as an extractant reduces the range of organo-halogen compounds compared to the adsorption on active carbon; the more polar compounds will have low recoveries. This means that EOX focuses on compounds that may be adsorbed on sediment and soil. EOX is used in the Netherlands as an indicator for contaminated soil.

Group parameters like AOX and EOX have to be strictly standardised. Especially changes in pre-treatment and isolation steps may affect the result for one compound but not for another, depending on the polarity. After isolation of the organo-halogen compounds all methods are using the same methodology for the determination of the organo-halogen content. An aliquot of the extract (1–100 µl) or the total amount of loaded active carbon (50–100 mg) is combusted at a temperature of at least 900°C. This converts the organo-halogens quantitatively into hydrogen halides. After passing through a scrubber filled with concentrated sulphuric acid to remove water from the gas stream, the hydrogen halides dissolve in glacial acetic acid or some other electrolyte present in the titration cell. The halides react with silver ions in the titration cell to form insoluble silver halides. The decrease in silver ions is measured by a silver indicator electrode and silver ions are added to restore the original concentration. This argentometric set point titration is normally done by micro-coulometry, i.e. an electric current will release silver ions from a silver electrode. The amount of electrical charge needed for restoring the set point is proportional to the number of moles of hydrogen halides. Assuming that all halides are chlorides, the number of moles is converted into a mass and results are expressed as e.g. 'micrograms per kilogram as chloride'.

There is no international standard for the determination of EOX or AOX in soil. In the Netherlands, NEN 5735 (1999) describes the determination of EOX in soil and sediment. In Germany, two standards for organo-halogen in sludge and sediment exists: DIN 38414 part 17 (1989) for POX and EOX and DIN 38414 part 18 (1989) for AOX. The latter standard uses an aqueous sodium nitrate solution to liberate organo-halogens from the sludge or sediment sample. Twenty milligrams of active carbon is added to the suspension and after filtration the active carbon including the original sample (<100 mg) is combusted. Organo-halogen compounds that were not liberated by nitrate will, therefore, also be combusted. Strictly speaking this method is therefore more than only AOX. Taking a representative sample of such a small size may already be difficult for sludge and sediment; for soil it requires an elaborate pre-treatment including drying and milling to a grain size of less than 100 µm. Such a pre-treatment is difficult to control regarding losses of more volatile compounds or contamination of the sample by the equipment and laboratory air.

EOX and AOX are useful indicators. As with most indicators there is also a need for further investigations if relatively high contents of EOX or AOX are found in a sample. Using direct GC–MS as the next step often gives disappointing results. Especially with EOX better results will be obtained if the next step consists of a clean-up over alumina or silica, after which the EOX content is measured again. In many cases the EOX content after clean-up is much lower, indicating that a large part of the EOX was caused by relatively polar substances and these substances are less amenable for GC–MS measurement. High EOX contents after clean-up indicate that large amounts of non-polar organo-halogen compounds are present and GC–MS may be used to identify these compounds.

The breakdown of high AOX values is even more complicated. There is no extract left for further investigation because all of the active carbon is combusted and the polarity range of AOX is much larger than GC–MS. This is a good argument to prefer EOX over AOX for soil analysis.

8.6.2   Phthalates

Phthalates, the di-esters of *o*-phthalic acid, are ubiquitous in the environment. They are used as plasticisers in different kinds of plastics. One of the most widespread phthalates, di-(2-ethylhexyl)-phthalate or DEHP, occurs in the list of priority pollutants (Annex X) of the Water Framework Directive of the European Union. In the coming years measures have to be taken to reduce discharges, emissions and losses of DEHP into the environment.

Due to the widespread use phthalates can be found everywhere, not only in soil but also in the laboratory environment, in the usual packing materials for samples and also in solvents. Controlling and reducing the blank values therefore is the main problem in the determination of phthalates.

With the exception of dimethyl-, diethyl- and dipropylphthalate, all the other phthalates can be considered as being non-polar. They can be easily extracted with a method for non-polar compounds. The recovery of dimethyl-, diethyl- and dipropylphthalate will be poor, because they will be partly removed during clean-up. If these compounds have to be analysed, an extraction as described for herbicides will be more effective. Phthalates tend to adsorb on glass surfaces if they are dissolved in non-polar solvents. Great care should be taken that phthalates are neither introduced nor lost during the determination.

Although electron capture detection is sufficiently sensitive for phthalates, the preferred method of determination is GC–MS. Phthalates give easily recognised mass spectra; with the exception of dimethylphthalate all phthalates show the characteristic fragment ion with m/z 149 as the base peak.

There is no standardised method for the determination of phthalates in the soil. Currently, a standard for the determination of selected phthalates in water (ISO/DIS 18856 (2002)) is in preparation within ISO/TC 147 – Water Quality.

The method is also applicable to soil, provided that a suitable procedure for the extraction of phthalates from soil can be derived. The scope of ISO 18856 contains the following phthalates: dimethylphthalate (DMP), diethylphthalate (DEP), dipropylphthalate (DPP), di-(2-methyl-propyl)phthalate (DiBP), dibutyl-phthalate (DBP), butylbenzylphthalate (BBzP), dicyclohexylphthalate (DCHP), di-(2-ethylhexyl)phthalate (DEHP), dioctylphthalate (DOP), didecylphthalate (DDcP) and diundecylphthalate (DUP); a further 38 phthalates are mentioned in an annex.

Solid phase extraction is used for the extraction of the phthalates from water samples. This minimises the use of solvents and reduces in this way the blank values. ISO 18856 pays much attention to the cleaning of the laboratory glassware. All glassware is heated to 400°C in a programmed temperature muffle furnace. This removes the phthalates but it also activates the glass surface. Deactivation of the inner surface is done by rinsing with iso-octane. Sample bottles are treated in the same way and stored in aluminium foil or appropriate stainless steel containers.

In principle, the extraction of phthalates from soil is rather straightforward; phthalates easily dissolve in both acetone and hydrocarbon solvents. The practical application requires much more attention than with other compounds of low polarity in order to reduce and control the blank. Good results can be obtained by the shaking procedure, using screw capped centrifuge tubes of 50 or 100 ml capacity. These tubes can be placed on a mechanical shaker and after short centrifuging the extract can be easily separated from the soil sample. In this way both the amount and the surface area of the glassware are minimised. Depending on the type of sample, a clean-up step may be necessary. ISO 18856 describes a conventional clean-up over a short column filled with alumina.

Care should be taken during concentration of the hydrocarbon solvent extract. Although phthalates have low polarity they tend to adsorb on glass surfaces if dissolved in non-polar solvents. This means that a rotary film evaporator should be avoided for concentrating the extracts; small size Kuderna-Danish concentrators are a better choice. If nitrogen gas is used for the final concentration, care should be taken to avoid any plastic tubing. The use of conventional chromatographic copper or stainless steel tubing offers sufficient flexibility and can be cleaned with acetone. A solid phase cartridge in the final gas supply will remove traces of phthalates from the nitrogen gas.

Interpreting the results of phthalate determinations in soil may be rather difficult. Even if blank values are really under control, phthalate contents, especially DEHP, may be caused by some artefacts, even the presence of pieces of plastics in the sample. If large samples have been pre-treated by milling, such an artefact may be difficult to recognise. Therefore, a different strategy should be considered and taking a number of small size samples that can be visually inspected may be a better choice. In any case the analyst should have a close look at the contents of the sample container in taking the required portion for the determination, and report any observations about the visible differences from normal soil.

8.6.3 Organotin compounds

Organotin compounds are organo-metallic compounds that consist of a single central tetravalent tin atom, covalently bound to one or more non-polar organic groups such as alkyl- and aryl-groups, but also esters or mercaptides. Tri-organotin compounds are used because of their strong fungicidal and biocidal activities. They are used as antifouling paints on ships and for wood preservation (tributyltin), as fungicide (triphenyltin) and as acaride (tricyclohexyltin and fenbutatin). Mono- and di-organotin compounds are used as heat and light stabilisers in polyvinyl chloride (PVC). In soil they are normally applied as acetate for crop protection, but they quickly hydrolyse to hydroxide. These compounds strongly adsorb to soil or sediment particles. In the environment they decompose (UV-light, biological and or chemical reaction) to di- and mono-organotin and non-toxic tin(IV) compounds (Brandsch 2001).

Depending on the number of organic groups, tetra-, tri-, di- and mono-organotin compounds can be distinguished. If less than four organic groups are present, the remaining places on the tin atom are occupied by anions like chloride or hydroxide. This gives the tri-, di- and mono-organotin compounds an increasing polar character. For the extraction from soil, it means that tri-organotin compounds can be considered as non-polar compounds (see also Table 8.1) and can be extracted with several non-polar solvents (e.g. hexane). The same solvents give low or no recoveries when applied to the more polar di- and mono-organotin compounds (Abalos *et al.* 1997; Arnold *et al.* 1998). They are much too polar and in order to extract the mono- and di-compounds, it is necessary to reduce the polarity of the compounds. The extraction efficiency can be improved by adding a reagent with ion pairing properties (HCl, HBr or HAc are frequently used). Tropolone (2-hydoxy-2,4,6-cycloheptatrien-1-one) is often used to increase the extraction efficiency of mono- and dialkyl-organotin compounds. It is added to the non-polar extraction solvent. Mono- and dialkyl-organotin compounds form stable complexes with tropolone, that will dissolve in the extractant, increasing the extraction efficiency. If the organotin is not accessible by the solvent, tropolone has no effect.

For water it is possible to devise a method that is able to extract all the organotin compounds. Most methods for soil are based on these well-established methods. In water, the partition between the sample and the solvent will always be in favour of the solvent. By using HBr or HAc, the polarity is further reduced, which improves the extraction efficiency of the mono- and di-organotins. The same procedure applied in soil will also reduce the polarity of the organotins. As long as the extractant is strong enough, the organotins will go to the organic solvent. The advantage is that the polar organotins become more readily extractable, as long as they are accessible for (physically available to) the solvent. However, due to the addition of water, the non-polar organotin compounds become more difficult to extract, because their accessibility is reduced. The effect of broadening the extraction method is the reduction of the extraction efficiency of part of the target compounds from soil. Application of water methods to soil does not automatically lead to a robust method, because the soil properties in combination with the organotin properties influence the results. The properties of all relevant organotin compounds may differ too much to extract them with one single procedure. In some standard methods general procedures for organotin compounds are described, but validation is still difficult. In the Dutch draft-standard (NVN 5729: 2001) acidification and tropolone are both used in the extraction procedure. After the separation of soil and extraction liquid, the extracted organotin compounds are derivatised and measured. The method should be applicable for all organotin compounds, but the results of an intra-laboratory were poor. In Germany another approach is under investigation. The derivatisation reagent is added to a mixture of the soil and an acetate buffer (pH 4.5), before extraction. The derivatised organotin is extracted with hexane. With this single step procedure, simultaneous derivatisation and extraction, the extractability of the derivatised organotin compounds is probably better, which is a great advantage. In a preliminary research study, recoveries of ca. 80% are reported for monobutyltin (Ad-hoc Arbeitskreis 2002). A disadvantage can be that other soil compounds are also derivatised, which requires extra reagent. These other derivatised compounds may give interference peaks in the chromatogram. Further research is necessary to obtain a robust method.

By reduction of the scope of the investigation to the non-polar (in general more toxic) organotins and accepting a relative high detection limit, a simple robust method can be used. If the concentration is high enough, the extraction can be followed by a relatively simple measurement. Compano *et al.* (1995) described an extraction with methanol for triphenyltin and tributyltin followed by HPLC-separation, post-column derivatisation and fluorimetric detection.

For measuring at low concentrations, the derivatisation of the extracted organotin compound with sodium tetraethylborate is an established procedure to make a more sensitive gas chromatographic analysis possible. Derivatisation can be combined with the extraction as prescribed in the German method but can also be carried out after the extraction. Selective detection is possible using GC–MS. For organo-metallic compounds atomic emission or atomic absorption detection is not only very sensitive, but very selective. The steps necessary for the chromatographic separation and detection are described in ISO/CD 17353 (2001). Application of this well-described method for soil samples does not mean that the difficulties with the extraction have been overcome.

As shown in this sub-chapter, the developments in the analysis of all the organotin compounds in soil has not led to a well established and general accepted analytical method. This is because the target compounds differ too much in polarity. Reducing the scope of the analytical method decreases the necessary steps of the analytical procedure, resulting in a more robust system The most reliable and robust method will be obtained when the analytical scope has been reduced to a selection (based on the properties of the organotin compounds). More research is necessary for the development of a method with a broad scope.

8.6.4 More polar herbicides and pesticides

Herbicides are components with relatively low K_{ow} values and high solubilities. This means that they can be extracted from soil with a water/acetone mixture. To enable the measurement of low concentrations, the herbicides have to be transferred to a volatile solvent, which is not miscible with water, like petroleum ether or dichloromethane. If the original acetone/water extraction medium is concentrated (evaporation), the water percentage will increase during the concentration and the extracted components will precipitate or adsorb to the equipment. It is therefore necessary to add a solvent like petroleum ether to the system. There will be a distribution between the soil and the acetone/water mixture and one between the acetone/water mixture and the petroleum ether. If the water content is low the system is very effective for non-polar components. With a higher water content it will be more effective for more polar components. With a too high water content the accessibility increasing properties of acetone are lost and the efficiency of the extraction process decreases.

After the removal of the soil the acetone can be removed by extraction with water. Non-polar components will stay in the petroleum ether phase. More polar components like the herbicides may be removed together with the acetone from the petroleum ether.

The ratio between the selected solvents is therefore critical. In the method for herbicides in soils, described in ISO/DIS 11264 (2002) (derived from Steinwandter 1991) the first step in analyses is the estimation of the water content. An amount of water is added to the aliquot of the field moist sample to give a total of 100 ml of water. Then 200 ml of acetone is added and the sample is shaken for 6 hours. The extracted herbicides are transferred to an organic phase by adding 150 ml of petroleum ether or DCM and shaking for another 5 minutes. The critical solvent ratio used in this procedure is water:acetone:DCM is 1:2:1.5. The organic layer is chemically dried and evaporated to dryness. The residue is dissolved in acetonitrile and analysed by HPLC. In this procedure DCM gives the highest recovery. The recoveries with petroleum ether are slightly lower (up to 10%) for some soils and analytes. This must be appreciated when the use of DCM is restricted for health and environmental reasons. In the

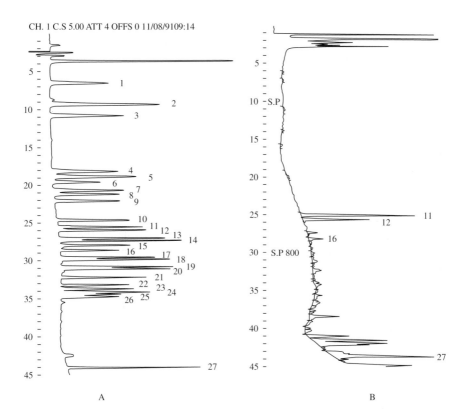

No.	Compound	No.	Compound	No.	Compound
1	Atrazine-desisopropyl	10	Methabenzthiazuron	19	Dichlobenil
2	Metamitron	11	Chlortoluron	20	Terbuthylazine
3	Atrazine-desethyl	12	Atrazine	21	Chloroxuron
4	Hexazinon	13	Isoproturon	22	Propyzamide
5	Metoxuron	14	Diuron	23	Terbutryn
6	Bromacil	15	Metobromuron	24	Ethofumesate
7	Simazine	16	Metazachlor	25	Metolachlor
8	Monuron	17	Sebuthylazine	26	Alachlor
9	Cyanazine	18	Propazine	27	Pendimethalin

Figure 8.4 Chromatogram of herbicides. A, standard solution; B, Chromatogram of a soil sample. Stationary phase: C18 Column, Mobile phase: water and acetonitrile with gradient programme, Injection volume: 50 µl, Oven temperature: 30°C, Flow rate: 1.0 ml/min, Detection wavelength: 235 nm (Chromatograms are supplied by G. Offenbächer).

presented solvent ratio, the recoveries using ethyl acetate or diethyl ether instead of DCM are significantly less. In the HPLC procedure a programmable UV-detector has to be used and chromatograms as presented in Figure 8.4 can be obtained.

8.6.5   Phenols

Phenols are very polar (phenol and dihydric phenols) to very non-polar (nonylphenol) as shown in Table 8.1. This makes it very complex to analyse the phenols in a single GC run. The non-polar phenols can be extracted with an acetone–hydrocarbon system as used for PAHs and PCBs. Using this extraction system may be effective for phenol, but during the washing step with water for removal of the acetone the phenol will be discharged with the water phase. Successful analyses of phenols in soil will be achieved if the scope is limited to the phenols that can be extracted and analysed in one run.

(a)   Polar phenols ($K_{ow} < 2.5$). These compounds have some acidic charac-
       ter and can be extracted from soil with water at a higher pH. For the
       following GC measurement it is necessary to make less polar derivatives
       as is prescribed for the analyses of phenols in water.

(b)   Non-polar phenols. Compounds such as octylphenol and nonylphenol
       may be extracted using a method developed for the analysis of PAHs
       or PCBs. After extraction they may be directly injected on a GC with
       FID or MSD. Because a lot of disturbing substances can be present,
       MSD detection is recommended. In spite of the high K_{ow} value these
       compounds have still a polar tail, which may be adsorbed on polar
       surfaces in the same way as phthalates (see Section 8.6.2). If the
       injection system of the GC is contaminated a compound like nonyl-
       phenol will be absorbed in the injector and will not be detected.
       A check on the performance of the chromatographic system is therefore
       necessary.

An intermediate polar compound like trimethylphenol can be analysed with both the systems. A lower recovery with the first method will be obtained if soil with a high organic matter has to be analysed. In that case the phenol is more strongly adsorbed to the soil matrix. The washing step in the second method will also reduce the recovery. An alternative is making use of the extraction procedure for the chlorophenols. This procedure makes use of the weakly acidic character of the phenols.

Polar phenols can be analysed by HPLC with UV-absorption (at about 280 nm) and/or UV-fluorescence detection (excitation at 230 nm; emission at 590 nm) with the same type of C18-reversed phase columns that are used for the determination of PAHs. Combined with the isolation of polar phenols from water by solid phase extraction a simple and straightforward method of analysis is possible. The extraction efficiency for phenol itself is strongly dependent on the type of solid phase material used for the isolation. Common C18 cartridges may show relatively low recoveries for phenol; with some types, for example, Lichrolut EN (Merck), recoveries better than 95% may be obtained.

There are different approaches for the isolation of polar phenols from the soil. An aqueous solution of potassium carbonate will extract the polar phenols as phenolate anions. However this extract may contain large amounts of humic substances as well, depending on the type of soil and the content of organic matter. In some cases, artefacts may be present due to the degradation of humic substances under these basic conditions. These effects restrict the application of the direct extraction at high pH. Under acidic conditions the polarity of the polar phenols is greatly reduced by the suppression of the dissociation. In this case the extraction with a low polarity solvent such as toluene or dichloromethane gives satisfactory results. Re-extraction of the organic phase with an aqueous potassium carbonate solution acts also as a clean-up i.e. non-polar compounds such as PAHs and PCBs remain in the organic phase and the phenolic compounds are transferred into the aqueous phase.

Although many manufacturers of HPLC-columns and SPE-cartridges have application notes (e.g. Merck, Chrompack) for the determination of polar phenols, a standardised method is not yet available.

References

Abalos, M., Bayonna, J.M., Compano, R., Granados, M., Leal, C. & Prat, M.D. (1997) Analytical procedures for the determination of organotin compounds in sediment and biota: a critical review. *Journal of Chromatography A*, **788**, 1–49.

Alexander, M., Chaney, R., Cunningham, S.J., Harmsen, J. & Hughes, J.B. (2003) Chemical measures of bioavailability. In R. Lanno (ed.), Contaminated soils: from soil–chemical interaction to ecosystem management. (SETAC-publication in press.)

Arnold, C.G., Berg, M., Müller, S.R., Dammann, U. & Schwarzenbach, R.P. (1998) Determination of organotin compounds In water, sediments and sewage sludge using perdeuterated internal standards, accelerated solvent extraction, and large volume-injection GC/MS. *Anal. Chem.*, **70**, 3094–3101.

Ad-hoc Arbeitskreis (2002) im DIN NAW I3 UA4 "Zinnorganische Verbindungen in Schlämmen und Sedimenten".

Brandsch, R. (2001) Risikobewertung für eine Landablagerung von Tribytylzinn-kontaminierten Hafensediment: Structur-Wirkungsbetrachtungen und Mechanismen des biologischen Abbaus. Ph.D. thesis, University Bremen, Germany.

Compano, R., Granados, M., Leal, C. & Prat, M.D. (1995) Liquid chromatographic determination of triphenyltin and tributyltin using fluorimetric detection. *Analytica Chimica Acta*, **314**, 175–182.

DIN 38414 Part 17 (November 1989) Deutsche Einheitsverfahren zur Wasser-, Abwasser und Schlam- muntersuchung. Schlamm und Sedimente – Bestimmung von ausblasbaren und extrahierbaren, organisch gebundenen Halogenen. (German standard methods for investigation of water, waste-water and sludge. Sludge and sediments – Determination of purgeable and extractable organically bound halogens.)

DIN 38414 Part 18 (1989) Deutsche Einheitsverfahren zur Wasser-, Abwasser und Schlammunter-suchung. Schlamm und Sedimente – Bestimmung von adsorbierten, organisch gebundenen Halo-genen. (German standard methods for investigation of water, wastewater and sludge. Sludge and sediments – Determination of adsorbed organically bound halogens.)

EPA Method 3545A (1998) Pressurized fluid extraction (PFE).

European Commission (2002) Implementating Council Derective 96/23/EC concerning the performance of analytical methods and the interpretation of results. *Official Journal of the European Communities*, 17.8.2002. L221/8–L221/36.

ISO/DIS 22892 – Soil Quality – Guideline for the GC/MS identification of target compounds.

ISO/DIS 18856 (2002) Water quality – Determination of selected phthalates using gas chromato-graphy/mass spectrometry.

ISO/CD 17353 (2001) Water quality – Determination of selected organotin compounds – Gas chroma-tographic method.

ISO/DIS 11264 (2002) Soil Quality – Determination of herbicides using HPLC with UV detection.

ISO 13877 (1998) Soil Quality – Determination of polynuclear aromatic hydrocarbons – Method using high-performance liquid chromatography.

ISO/FOIS/14507 (2002) Soil Quality – Pre-treatment of samples for the determination of organic contaminants in soil.

NEN 5735 (1999) Bodem – Bepaling van het halogeengehalte afkomstig van niet-vluchtige, met aceton en petroleumether extraheerbare organohalogeenverbindingen (EOX). (Soil; Determination of the halogen content originating from non-volatile, with acetone and petroleum ether extractable organohalogen compounds (EOX)).

Ontwerp-NVN 5729 (2001) Bodem – Bepaling van organotinverbindingen in grond en sediment. (Preliminary Standard. Soil – Determination of organotin compounds in soil and sediment).

Reid, B.J., Jones, K.C. & Semple, K.T. (2000) Bioavailability of persistent organic pollutants in soils and sediments: a perspective on mechanisms, consequences and assessment. *Environmental Pollution*, **108**, 103–112.

Rulkens, W.H. (1992) Soil remediation using extraction and classification (in Dutch). *Procestechnologie*, **9**, 43–55.

Steinwandter, H. (1991) A new extraction principle of the micro on-line method using the binary solvent system water + acetone. *Fresenius J. Anal. Chem.*, **342**, 150–153.

Tomlin, C.D.S. (ed.) (2000) The Pesticide Manual. British Crop Protection Council, Farnham, UK.

Traag, W.A., van Leeuwen, M., van Galen, F.R. & Harmsen, J. (1997) Development and limited validation of methods for pre-treatment and extraction for the determination of organic contaminants in solid matrices with gas chromatography-mass spectrometry (in Dutch). RIKILT report 97.27, Wageningen, The Netherlands.

9 Leaching tests

Leslie Heasman

9.1 Introduction

Leaching tests are designed to provide a means of measuring the overall release of soluble constituents from a solid matrix. Leaching tests provide a direct measurement of the combined effect of the processes that are releasing contaminants and they eliminate reliance on models which are based on poorly understood processes. The design of the leaching test and the selection of the appropriate leaching test for the particular assessment are, of course, crucial to the reliability and relevance of the results. Leaching tests are empirical and, unless the leaching protocol is followed exactly, the value of the data produced will be questionable.

The factors which control the migration of contaminants from a solid matrix into solution and their interactions are many and complex. The actual rate of leaching of a single contaminant is not related simply to solubility. Contaminated soil is a complex system comprising multiple contaminants which are present as a number of species in a mixture of heterogeneous matrices.

Most decisions regarding the potential impact on the environment or human health of contaminants in soils are based on an assessment of the total concentration of contaminants present in the soil. It is, however, recognised that the proportion of contaminants that will be released to the environment generally will be less than the total concentration of contaminants present in the soil.

As our understanding of risk assessment and the natural attenuation of contaminants increases, it is important that we are able to obtain reliable data from which can be quantified the probable release into the environment of contaminants from a particular material in a particular set of circumstances. The availability of this data will allow a more robust assessment of the environmental impact of that material in the specified circumstances.

The mobilisation of contaminants and their rate of release from a solid form or when contained within the solid structure of a soil matrix will depend on a number of specific factors relating to the soil matrix, the characteristics of the contaminants and the environment in which the soil is present. It is necessary to be able to estimate the concentration and/or flux of contaminants that will be released and to estimate the changes in concentration or flux over time.

Many computer-based models, e.g. Parkhurst and Appelo (1999), have been developed to calculate the movement of soluble contaminants from and through soils. Models rely on assumptions regarding the processes that take place, the equations used to calculate the effect of those processes and the values that are used for each parameter. The reliability of the output from a computer model will depend on how well the situation being modelled is understood, on the complexity and thoroughness of the model, the assumptions made and the values selected for the variables. The relevance of the output will depend on whether the model and all the assumptions are appropriate to the situation being assessed. None of the models available provide suitably robust and meaningful outputs, hence there is a need for leaching tests that are relatively short, reproducible and cost-effective.

9.2   The value of leaching tests in land contamination

Leaching tests play a valuable role in the assessment of land contamination for a number of different purposes. Data on the flux of contaminants from soil, over time, can be used directly in the assessment of risk to groundwater and surface water. Leaching tests, involving sequential extraction, designed to mimic the acid digestion mechanisms in the stomach can be used to assess the availability of contaminants for absorption into the body (Kelley *et al.* 2002). Bioavailability leaching tests are not covered in this chapter but the principles discussed for leaching tests are also relevant to bioavailability leaching tests (Ruby *et al.* 1996; Rodriguez *et al.* 1999) see also Chapter 1 (1.9.3).

On-site treatment may be proposed if it is determined that the release of contaminants presents an unacceptable risk. Leaching tests can be carried out on the treated soil to monitor whether the treatment has achieved the intended reduction in contaminant mobility.

If it is necessary to remove contaminated soil from the site for disposal at a permitted landfill, it will be necessary to determine whether the soil will meet the acceptance criteria for a non-hazardous or hazardous site. If the soil does not meet the acceptance criteria, treatment will be necessary to remove the mobile contaminants or to reduce their mobility.

If the site is located in an ecologically sensitive environment and it is necessary to determine the ecotoxicity of the water which percolates through the contaminated soil, leaching tests can be used to provide the eluate for use in the ecotoxicity tests (see Chapter 10).

9.3   The primary factors which control leaching

There are a number of factors which control the rate of release of contaminants into solution. These include:

1. The leaching mechanism (predominantly kinetically or diffusion controlled).
2. The chemistry of the contaminants of concern.
3. The pH and reduction/oxidation (redox) potential of the environment in which the material is present.
4. The nature of the leachant.
5. Physical, chemical or biological changes in the material over time.
6. The temperature at which the leaching test is carried out.
7. The duration of the test.
8. The nature of any agitation used.

9.3.1   The leaching mechanism

The release of contaminants from a solid into solution involves a number of interrelated mechanisms. Dissolution from solid to liquid occurs at the interface between the particle surface and the liquid passing over it. The overall surface area to mass or volume ratio and the intimacy of contact between the particle surface area and the liquid depend on the particle size and shape distribution and the pore structure of the solid matrix.

Leaching is controlled by the rate at which liquid flows past individual particles. For situations and materials where the liquid flows around the particles in the solid matrix, the rate of flow is driven by the hydraulic head of the liquid above the solid material and the efficiency of solute transfer is controlled by the contact between the surface area of the particles and the liquid. Where the rate of flow is very fast and the time over which a state of solute transfer equilibrium is established is slow, leaching may not reach true chemical equilibrium hence not all the leachable components will be leached as quickly as in the situation where the rate of flow is slower. Leaching tests are all based on the assumption that equilibrium or semi-equilibrium is reached in the timescale of the test.

For situations where there is little or no flow through the solid matrix, for example where the material has an exceptionally low hydraulic conductivity compared with the surrounding material, the liquid flow will be predominantly around large blocks or masses of the material. In such situations, solute transfer is affected by the diffusion of contaminants from inside the bulk of the material body to the surface of the body thence into solution in the liquid passing over the structure.

For the purposes of describing leaching tests, the description of the leaching mechanisms hence the description of the types of tests are simplified into those which are relevant to granular materials and those which are suitable for monolithic materials.

In leaching tests for *granular materials*, it is assumed that the liquid percolates through a material which comprises a mass of particles or granules. As the liquid comes into contact with the surface of each particle, contaminants transfer into solution and migrate with the liquid. The mechanism for the movement of contaminants out of each individual particle is diffusion. The concentration of contaminants in the leachate is controlled by the relationship between the rate of liquid movement through the material and the rate of dissolution of contaminants into the leachate.

In leaching tests for *monolithic materials*, it is assumed that the permeability of the material is so low that diffusion is the predominant controlling factor, with the movement of the liquid mainly over the surface of a block of material rather than around individual particles or granules of material. Thus, other than initial surface wash off effects, the release of contaminants depends on the movement of contaminants by diffusion through the tortuous pathways of the interstitial spaces inside the block to the surface of the block where the contaminants pass into solution in the external liquid. Release is therefore related to the exposed surface area of the material. Tests carried out on ground or disaggregated material will not therefore reflect release in a *real* situation. Diffusion is generally the predominant leaching mechanism for materials such as cement stabilised and solidified materials provided that the matrix is well formed and has a low primary and secondary permeability.

9.3.2   The chemistry of the contaminants of concern

The chemistry of the individual contaminants and their compounds or complexes will affect the solubility of the contaminant under specific conditions. Species-specific solubility factors, rates of adsorption and desorption and the change in these factors in different pH and redox environments together with the effect of the nature of the leachant, all are important in controlling release from the solid matrix.

9.3.3   The pH and redox potential of the environment in which the material is present

Two of the most important factors which control the solubility of chemical compounds are the pH and the redox potential of the environment in which the material is present. Many contaminants exhibit a sharp change in solubility over small changes in pH or redox potential (van der Sloot *et al.* 1997). For all leaching tests and assessments of environmental impact, it is important to identify the critical zone in the pH value or redox potential and how it relates to the environment in which the material is located or into which it will be placed and how the pH and redox potential of that environment may change with time.

9.3.4   The nature of the leachant

The nature of the leachant clearly will affect directly the leachability of the contaminants depending particularly on the pH and redox potential of the liquid and the presence of chemical species that may affect the solubility of contaminants, in particular, the presence of organic compounds and complexing agents. The nature of the leachant used in different leaching tests usually has been chosen based on the purpose for which the test was developed. For example, the United States TCLP (USEPA 1990) test and the UK WRU Waste Research Unit test (Young & Wilson 1982) were prepared for use in the assessment of the leaching behaviour of materials placed in landfill sites. The leachants used in these tests are designed to replicate the significant properties of landfill leachate. Most tests specify the use of distilled or de-ionised water as it is widely available as a chemically consistent leachant and is similar to mildly acidic rainwater. If the leachant specified in the test selected is not appropriate for the environmental situation which is being assessed, consideration should be given to the use of an alternative, more representative leachant.

9.3.5 Physical, chemical or biological changes in the material over time

Leaching tests normally are carried out over a limited period of time, usually over 24 hours. The nature of the solid material will change only in response to reactions which occur in the time period and by the processes covered by the test. Medium and long-term changes to the nature of the material which can affect its leaching properties will not be recorded by the test. Such changes include those caused by biodegradation of the contaminants present and of the matrix in which the contaminants are held. The process of biodegradation may also significantly change the pH, redox potential and chemical nature of the environment in which the material is located. The process of leaching may itself, over time, result in changes to the physical structure of the matrix in which the contaminants are held leading to a change in retention and release characteristics. For materials which have been stabilised, for example using cement or lime, the full benefit of the hydration process may not be achieved for some time due to the progressive curing which takes place. Conversely, for some stabilised materials, long-term leaching and the effect of carbonisation from atmospheric carbon dioxide or acidic leachants degrade the physical structure of the matrix. The potential for such changes must be taken into account in the assessment of the rate of contaminant release and environmental impact.

9.4 Available leaching tests

A plethora of leaching tests has been developed, many of which were developed or are used based on a poor understanding of the controlling factors and leaching processes. The tests have been developed for many different purposes and most have been developed in isolation from each other. In effect, the tests represent numerous slightly different ways of measuring the same thing – the release of contaminants into a liquid passing through or round a solid material. Due to the variety of tests available the results obtained from different tests rarely are directly comparable.

9.4.1 Types of leaching tests

The leaching tests available can be divided into those which are intended to test the effect of a specific factor such as pH or redox on leaching, and those such as batch tests, column tests and field tests which are intended to replicate the overall combined effect of all relevant factors. Alternatively, the tests can be divided according to the level of detail which they provide. The latter division is used by CEN Technical Committee, TC/292, which is working to develop standards for the characterisation of wastes including a number of leaching tests. The leaching standards being developed by CEN TC/292 are directly relevant to the assessment of leaching from non-waste materials as the definition of waste is a legal definition dependent on whether someone wants the material or not and is not a definition based on the chemical, physical or biological nature of the material. Similar to contaminated soil, wastes commonly comprise complex heterogeneous mixtures of multiple contaminants. The tests are divided by CEN TC/292 as follows:

1. *Basic characterisation tests* that are used to obtain information on the short and long-term leaching behaviour and characteristic properties of materials.
2. *Compliance tests* that are used to determine whether the material complies with specific reference values such as soil screening levels for remediation purposes, or waste acceptance criteria for landfill disposal. These tests focus on key variables and aspects of leaching behaviour identified by basic characterisation tests.
3. *Verification tests* that are used as a rapid check to confirm that the material is the same as that which has been subjected to the compliance tests.

9.4.2 Basic characterisation tests

The basic characterisation tests include percolation or flow through column, tests and multiple batch tests. Column tests comprise passing the leachant through a column filled with the solid material. It is important that the way in which the column is packed is as representative as possible of the material in the environmental situation being assessed. Column tests can be affected significantly by edge effects where leachant travels preferentially through fissures in the column and along the boundary between the column and the packed material. Generally, the larger the column the less significant the edge effects. Column tests are designed based either on the downward flow of the leachant under a fixed head or on the upward flow of leachant under a fixed pressure. The upward flow tests generally achieve a more even flow of leachant through the column. Leachate flowing from the column is collected in batches for analysis. The results represent the leaching pattern of the contaminants over time.

Multiple batch tests comprise the leaching of a single sample of material sequentially with fresh leachant, to identify the changes in leaching over time rather than the average rate of release of contaminants as in a single batch test. A low liquid to solid ratio (L/S ratio) such as 0.1 or 0.2 l/kg generally is used for each extraction in multiple batch tests. The low liquid to solid ratio for each extraction is commonly referred to as a bed volume. This is effectively the minimum L/S ratio necessary to conduct a shake test.

Basic characterisation tests include tests designed to assess the effect of a single variable on the leaching behaviour of a material. Such tests commonly comprise tests to establish the effect of leaching at different pH values and at different redox potentials.

9.4.3 Compliance tests

Compliance tests should be carried out based on available information on the main factors which control leaching from the material under consideration in the environment which is being assessed. Compliance tests generally comprise single batch leaching tests where a sample of the material being assessed is shaken for a specified time period with a specified volume of leachant. The main variables in compliance tests are the liquid to solid (L/S) ratio, the time of shaking and the leachant used. The L/S ratio specified commonly (CEN PrEN12457; NRA (1994) R&D Note 301; DIN 38414; AFNOR X 31-210; EPA TCLP) is 10:1 which is chosen based on practical issues such as ease of shaking and the provision of an adequate volume of leachate for analysis rather than on any meaningful measure of reality. The results obtained from a test using an L/S ratio of 10 represent the total release as a result of ten bed volumes of liquid passing through the material *in situ*. Depending on the rate of flow of liquid through the material in the environment, this can represent a very long time period. Single batch tests which are based on a smaller L/S ratio such as 2 represent release over a shorter time period.

9.4.4 Verification tests

Verification tests generally comprise tests *other* than leaching tests. They usually comprise checks on pH, appearance, odour and occasionally analysis of the *total* concentration of contaminants in a material. They may comprise shortened compliance type tests. There are no standard verification type tests that are used widely but CEN is developing such tests.

9.4.5 *Reviews of available tests*

A number of reviews of available leaching tests have been carried out (Environment Canada 1990; Wallis *et al.* 1992; Ure *et al.* 1993; Quevauviller *et al.* 1995; CEN/TC/292/NNI 1994; Science of the Total Environmental Special Issue 1996; van der Sloot *et al.* 1997). Examples of leaching tests performed throughout the world are shown in Table 9.1. As shown in the table, there are more tests available for granular materials than there are for monolithic materials.

Work is ongoing on the development of new tests and the improvement of existing tests. There are proposals to the European Union (EU) for projects to harmonise the existing tests and to narrow the bewildering range of tests available. A website was developed as part of the EU funded Network on the Harmonization of Leaching and Extraction Tests (www.leaching.net) which is updated as funds allow. Most of the development work is being carried out by, or is being followed by, CEN Technical Committee TC/292. The secretariat of TC/292 is held by the Standards Institution of the Netherlands, NNI.

9.5 Quality assurance and quality control

Some of the standard tests have been validated to assess the repeatability, reproducibility, reliability and robustness of the tests when performed in different laboratories and by different people in the same laboratories. Extensive validation work has been carried out for the USEPA tests and the CEN test PrEN 12457. Little validation work has been carried out to determine the reliability of the results from laboratory tests compared with the results from field tests. Some comparative work has been done between the results obtained from batch tests and the results from column tests (Young & Wilson 1982). In general, as the subject area is relatively new and knowledge is increasing all the time, the results of the tests must be interpreted in context. It is accepted generally that provided the medium to long-term potential changes to the environment and the material are considered and taken into account, the results obtained from leaching tests, particularly from aggressive shake tests, will be conservative compared with the results obtained from leaching *in situ*.

The method of sample pre-treatment prior to being subjected to leaching tests should be considered with care and taken into account in the assessment of results. As discussed in Section 9.3.1, tests which are carried out on ground or disaggregated material will not reflect the actual behaviour if the flow of leachant *in situ* is around rather than through the mass of the material. Similarly, pre-treatment such as drying will reduce the concentrations of any volatile component in the material and may change the chemical nature and hence the leaching potential of non-volatile components.

There are proposals submitted to collate available data from field studies as well as laboratory studies and to carry out some form of data comparison. It

Table 9.1 Examples of leaching tests used

Leaching tests for granular materials

Single batch leaching tests (equilibrium based)*				Multiple batch and percolation/column tests†				
				Serial batch tests				
pH domain 4-5	pH 5-6	Distilled or deionised water	Low L/S ratio	Low L/S	L/S > 10	Percolation/ column tests	Static methods	Speciation methods
TCLP	Swiss TVA	DIN 38414 S4	MBLP	UH Hamburg	NF-X31-210	NEN 7343 – Column up	Canada MCC 1	Sequential chemical extraction pH static test procedures (being developed by CEN TC/292)
EPTox		AFNOR X-31-210	CEN PrEN12457	WRU	WRU	ASTM – Column up	Canada MCC 2	
NEN 7341		Ö-Norm S2072	Wisconsin SLT		ASTM D4793-88	Column	Compacted granular tank leaching tests (Rutgers/ECN)	
California WET		CEN PrEN12457			NEN 7349	Germany – pH static		
Ontario LEP		Canada EE			MEP method 1320			
Quebec QRsQ		Canada MCC – 3C			Sweden ENA			
Soil – HAc		ASTM D 3987			MWEP			
		Soil – NaNO$_3$						
		Soil – CaCl$_2$						
		UK NRA test						

Leaching tests for monolithic materials‡

ANS 16.1
Tank leach test NEN 7345
Swedish MULP
Compliance and basic characterisation tank tests are being developed by CEN TC/292

Adapted from Table 2.1 in van der Sloot, et al. (1997) and reference should be made to this book for furthur details of the tests cited above.
*Compliance tests.
†Basic characterisation tests.
‡Dynamic leaching tests for both basic characterisation and compliance tests.

would also be of value to validate available computer solute transport models, including unsaturated flow models, with the results of laboratory leaching tests. Currently, however, extensive studies have not been carried out, hence broad validation of the results from computer models or laboratory tests against *in situ* data has not been possible.

As with other fields of contaminated soil testing, the lack of standard reference materials results in the lack of ready means of having the performance of leaching tests accredited as part of a laboratory test accreditation procedure. This is potentially problematic as more and more legislation, including particularly European Directives such as the Landfill Directive, specify that testing must be carried out by competent and suitably accredited laboratories.

It is important to recognise that the validation work which has been carried out for some of the tests (CEN, WRU, NRA) has been for leached inorganic parameters only. There is little data on which to determine whether any of the leaching tests are valid for use in assessing the leaching of organic contaminants. Factors such as volatilisation and absorption on to the test equipment of organic compounds have been raised as possible confounding factors but their impact has not been assessed to date.

9.6   Selection of a leaching test

Before deciding which leaching test is appropriate in a particular set of circumstances, information should be sought on the material under consideration. Where data is already available on analogous material it can be used to predict leaching behaviour. Knowledge of the chemistry of the material, the leachant and the environment in which it is located or into which it is being placed will facilitate an assessment of the probable sensitivity of leaching to parameters such as pH and redox and the importance of the presence of chelating agents, organic carbon or solubility controlling parameters in the leachant. It is important to understand whether the material should be tested as a granular material or as a monolithic material and whether equilibrium conditions or near equilibrium conditions will be reached during the test period.

Based on the information about the leaching behaviour of the material, the nature of the environment in which the material is or will be placed and an understanding of the nature and objective of the assessment, it is possible to select the most appropriate leaching test or battery of tests. It may be necessary to adapt existing tests in order to reflect more accurately the specific circumstances under consideration. The methodology behind the selection and adaptation of any tests used must be reported together with the results of testing.

Where a substantial amount of information is available regarding the leaching behaviour of a material in a specific set of circumstances, only a short, compliance type test will be necessary. On the other hand, for a material with poorly characterised leaching behaviour, more extensive basic characterisation testing will be necessary to obtain, with any confidence, the information on which to base an assessment of potential environmental impact. Given the range of leaching tests available, it is important that the aspects which will be and will not be identified by each type of test are well understood and are clearly identified in the scope of any such test and are included in the report on the results of the test.

Commonly, leaching tests are selected for use without full consideration of what they will and will not take into account. The most common factors considered in the selection of tests are convenience of application, cost, time to perform the test and minimisation of the number of leachate samples which must be analysed. Whilst these factors are very important considerations, they must be balanced against the risk of obtaining meaningless data which cannot be used correctly in an assessment of environmental impact.

The use of standard tests is helpful in many ways, as the results theoretically are comparable with tests carried out elsewhere using the same standard. The state of development of leaching tests is such that there is as yet no single accepted standard. In Europe, work is in hand to encourage the use, wherever possible, of the tests being developed by CEN TC/292. The TC/292 tests will be obligatory under the Landfill Directive for use in assessing whether waste materials can be accepted at certain types of landfill sites, hence it is likely that the standards will be specified more commonly by the regulatory bodies for other relevant applications.

Standards are not relevant in all circumstances. Standardised tests should not be used without considering whether they are appropriate to the particular set of circumstances under consideration. Where they are not appropriate, an alternative test or more appropriate leaching fluid should be selected and used and its selection in preference to the standard test justified.

9.7   Interpretation of the results of leaching tests

Most leaching tests are performed in a laboratory. Whilst leaching tests are designed and intended to reflect reality, there is a limited amount of data available from field tests which can be used to correlate with those from laboratory tests hence validate their performance in relation to the situation in the field. One of the most common errors in the interpretation of results from leaching tests is to assume that, based on the commonly used shaking test at a liquid to solid ratio of 10:1, the resultant leachant is representative of the concentration of contaminants that will emerge from the base of a deposit of the material tested. Even if it is assumed that the factors such as leachant used, pH and redox are correctly applied in the test, the concentrations of contaminants in the leachate represent an average of the concentration which will be leached out as ten bed volumes of leachant pass through the material. Without further understanding of the leaching behaviour of the material and its environment, it is not possible to tell whether most of the contaminants are flushed out as the first one or two volumes of leachant pass through the material or, alternatively, as the last volumes of leachant pass through the material following weathering or breakdown of a retentive matrix. This can give rise to errors of an order of magnitude in the estimation of the concentration of contaminants which will be present at the base of the material. Column tests or multiple batch leaching tests provide some indication of the changes in leaching behaviour with time.

Leaching tests on granular material can be used to determine leaching behaviour over an extended period of time. For example, a batch test using an L/S ratio of 10 which is equivalent to ten bed volumes can be related to time by determining the time which it will take for a single bed volume of leachant to pass through the material *in situ*. The time represented by the test at L/S 10 is ten times this time. Thus, batch tests and column tests carried out on granular material represent an acceleration of time. For materials in monolithic form, leaching depends predominantly on the diffusion of contaminants from the body of the material to the surface where it is transferred to the leachant. This takes place in real time and it is not possible to accelerate the process reliably. Tests carried out over time to measure leaching from monolithic materials provide time series data which can be extrapolated to determine the impact over longer time periods. Such extrapolations by their nature are based on the assumption that the rate of leaching remains stable and there are no changes to the controlling factors which will affect the leaching. There have been no very long-term tests carried out on monolithic materials to enable a comparison between extrapolated and actual long-term behaviour.

The interpretation of the results from leaching tests cannot be carried out without an understanding of the details of the test including those parameters which have been controlled or not controlled, an understanding of how the test conditions relate to reality and an understanding of how the conditions tested may change in the short, medium or long-term.

References

Characterisation of Waste in Europe (1994) State of the art report for CEN TC 292. STB/94/28, NNI 1994.

CEN/TC 292 PrEN12457-1 to PrEN12457-4 (2000) Characterisation of waste. Leaching compliance test for leaching of granular waste materials and sludges.

Environment Canada (1990) Compendium of waste leaching tests. Environmental Protection Series. Report EPS 3/HA/7.

Kelley, M.E., Brauning, S.E., Schoof, R.A. & Ruby, M.V. (2002) Assessing oral bioavailability of metals in soil. Battelle Press, Columbus. ISBN 157477123X.

National Rivers Authority (1994) Leaching Tests for Assessment of Contaminated Land. Interim NRA Guidance R&D Note 301.

Parkhurst, D.L. & Appelo, C.A.J. (1999) User's Guide to PHREEQC (Version 2) – A Computer Program for Speciation, Batch-Reaction, One-Dimensional Transport and Inverse Geochemical Calculations. U.S. Geological Survey Water-Resources Investigations Report 99-4259.

Quevauviller, P.H., Rauret, G., Ure, A., Rubio, R., Lopez-Sanchez, J.F., Fiedler, H. & Muntau, H. (1995) Certified reference materials for the quality control of EDTA and acetic acid extractable trace metals in soil. *Microchimica Acta*, **120**, 289–300.

Rodriguez, E.R., Basta, N.T., Casteel, S.W. & Pace, L.W. (1999) An in vitro gastrointestinal method to estimate bioavailable arsenic in contaminated soils and solid media. *Environ. Sci. Technol.*, **33**, 642–649.

Ruby, M.V., Davis, A., Schoof, R., Eberle, S. & Sellstone, M. (1996) Estimation of lead and arsenic bioavailability using a physiologically based extraction test. *Environ. Sci. Technol.*, **30**, 422–430.

Science of the Total Environment (1996) Special Issue on Leaching and Extraction Tests for Environmental Risk Assessment. Volume 178.

van der Sloot, H.A., Heasman, L. & Quevauviller, P.H. (1997) Harmonization of leaching/extraction tests. Studies in Environmental Science 70. Elsevier. ISBN 0–444–82808–7.

USEPA (1990) Part 261, Appendix II – Method 1311 Toxicity Characteristic Leaching Procedure (TCLP), *Federal Register*, **55**(61), March 29.

Ure, A., Quevauviller, P.H., Muntau, H. & Griepink, B. (1993) Report EUR 14763 EN, CEC. Brussels.

Wallis, S.M., Scott, P.E. & Waring, S. (1992) Review of leaching test protocols with a view to developing an accelerated anaerobic leaching test. AEA-EE-0392. Environment Safety Centre.

Young, P.J. & Wilson, D.C. (1982) Testing of Hazardous Wastes to Assess their Suitability for Landfill Disposal. AERE-R 10737, Environmental and Medical Sciences Division, AERE Harwell, Oxfordshire.

10 Ecological assessment and toxicity screening in contaminated land analysis

Andreas P. Loibner, Oliver H.J. Szolar, Rudolf Braun and Doris Hirmann

10.1 Introduction

Ecotoxicity can only be measured by the application of biological methods, whereas chemical analysis determines concentrations of defined chemicals that may be used to deduce toxic effects. All concentration levels such as screening values, guideline values, threshold concentrations, benchmark concentrations and trigger values used for the assessment of contaminated media (water, soil, sediments, etc.) should ideally be derived from the observation of biological effects. Comparing pollutant concentrations with these usually conservative standard values is a common practice in the preliminary assessment of contaminated sites. The integration of further information, such as ecotoxicity data from the site, can improve the risk assessment process and enable a more reliable prediction of environmental threats.

Toxicity testing comprises the evaluation of effects of known chemicals on ecological receptors (ecotoxicity tests), and the measurement of effects exhibited by contaminated media (*bioassays*). For ecotoxicity testing, key test organisms are exposed to a range of concentrations of a series of potentially toxic agents and the resulting adverse effects are measured. Besides the evaluation of chemical substances, toxicity tests have been used to measure toxic effects of contaminated water (surface water, wastewater, etc.), where in-depth information on the type and concentration of pollutants is missing. More recently, such tests have been developed for the evaluation of contaminated soils.

Data derived from comprehensive toxicity testing may be integrated in risk assessment frameworks for a more reliable estimation of the probability that the contaminated site poses environmental harm. Moreover, toxicity tests may be used as screening tools in order to identify polluted soil or water samples, and to reduce the number of samples that require full chemical and/or toxicological analysis.

10.1.1 Ecotoxicology

Early ecotoxicological investigations focused primarily on aquatic ecosystems and in particular on the impact of industrialisation on water quality. Acute toxicity tests indicated the effects of toxic chemicals present in industrial wastewater. After a standardised method was adopted by the American Society for Testing and Materials in the middle of the 20th century (Hoffman *et al.* 1995), it was suggested that the occurrence or non-occurrence of particular organisms in the aquatic environment is a more sensitive and reliable measure of ecosystem health than is the application of solely chemical or physical analysis. During recent decades, growing environmental awareness and increased experience in applying ecotoxicological assessments have resulted in an extensive range of toxicity test protocols focusing not only on aquatic, but also on terrestrial ecosystems.

The term *ecotoxicology* was described by Truhaut in the late 1960s for the very first time as comprising the result of a linkage between ecology, toxicology and chemistry (Forbes & Forbes 1994). Whereas ecology deals with interactions between different inhabitants of a distinct environment and consequences resulting from these interactions, the term *toxicology* has been defined as 'the study of adverse effects of chemicals on living organisms' (Klaassen & Eaton 1991). Gallo and Doull (1991) restricted the scope of toxicology to chemicals originating from anthropogenic activities. The term *xenobiotics* is used to describe substances that are not produced naturally and that do not occur as natural constituents in biological systems.

Hoffman *et al.* (1995) defined ecotoxicology as the science that forecasts effects of potential toxic substances on non-target species and ecosystems with the goal of protecting entire ecosystems and not only isolated components. Ecotoxicology does not cover toxicological investigations that deal exclusively with agricultural crops and domestic animals, or that focus strictly on human health. Ecotoxicology rather provides a scientific basis for the evaluation of environmental hazards and for supplying the information that can be used in making regulatory decisions.

10.1.2   Ecological risk assessment

Contrary to human health risk assessment that aims to protect the individual of a single species, the ecological approach focuses on the protection of multiple species or ecological functions. According to the USEPA (1992), ecological risk assessment is 'the process that evaluates the likelihood that adverse ecological effects may occur or are occurring as a result of exposure to one or more stressors'. Various procedures developed for ecological risk assessment address the same key elements: definition of the problem, quantification of exposure and effects, and characterisation of risk.

The scheme presented (Fig. 10.1) describes the subsequent steps of ecological risk assessment that are suggested by the USEPA (1998). The use

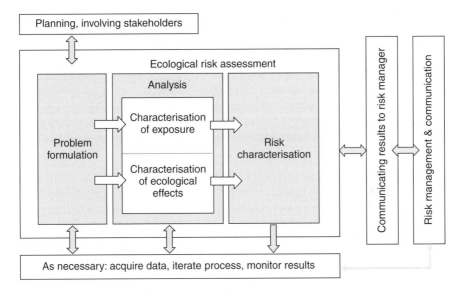

Figure 10.1 Framework for ecological risk assessment as defined by the USEPA (1998, redrawn).

of such frameworks for the assessment of contaminated sites is outlined briefly in the following passage:

Problem formulation. Within this process all available information about a contaminated site is collected including the nature of the contaminants and their sources, obvious effects and potential receptors as well as environmental recipients. Within this very first stage of the risk assessment procedure, an assessment endpoint has to be determined. Assessment endpoints are the expression of an environmental value (represented by an ecological entity) that is at risk, e.g. a distinct population that faces harm due to pollution. It has to be emphasised that toxicity-test endpoints or other measurement endpoints (in general, measured effects under test conditions) in most cases do not represent assessment endpoints (response of population or ecosystem in the field). Measurement endpoints should be representative for assessment endpoints or have a known relationship to the assessment endpoint allowing the extrapolation of data.

Characterisation of exposure. This process intends to estimate *how much* of a harmful chemical substance is *for how long* in contact with a specified organism. Exposure characterisation is usually a complex process that should consider physico-chemical properties of pollutants and the site (e.g. to estimate influencing factors such as the availability of contaminants), as well as naturally occurring pollutant degradation. For contaminated soils, exposure is usually estimated with measurement of toxicant concentrations by chemical analysis of site samples or biota and, if applicable, modelling the potential uptake of pollutants by the receptor organism.

Characterisation of effects. Assessment of effects comprises the determination of harmful effects as a function of exposure to a toxic chemical. Effects have to be related to the assessment endpoint. Effects may be derived from field studies conducted at the contaminated site (ecological epidemiology) or by performing toxicity tests with the chemical concerned. The latter are used more frequently in effect assessment.

Risk characterisation. Risk characterisation is a process that integrates results of the analysis phase (exposure and effects) and information pertaining to uncertainty of the estimation and description of risks. Within this phase, the likelihood of the current (diagnostic) or future occurrence (prognostic) of significant effects and their consequences on the assessment endpoint are estimated resulting in an appraisal of the severity and temporal extent of harm to the environmental value.

Risk management. This is a decision-making process that focuses on the minimisation of risks but is not part of the assessment procedure. However, communication between risk assessors and managers throughout the whole procedure is necessary for an efficient assessment. In the context of contaminated land assessment, a decision has to be taken whether remedial action is required or not.

Currently, the European Union is working on the standardisation of chemical risk assessment (White Paper on *Strategies for a Future Chemicals Policy*, European Commission 2001). Conceptual papers are already available, e.g. for the risk assessment of chemicals on terrestrial ecosystems (CSTEE 2000). Whilst these documents mainly focus on predictive risk assessment for the release of chemicals, the European Union's concerted action CARACAS published two volumes on risk assessment for contaminated sites comprising the scientific perspective (Ferguson *et al.* 1998) and a comprehensive description of national approaches of EU member states in assessing risks posed by existing contaminated sites (Ferguson & Kasamas 1999).

10.2   Toxicity testing

Toxicity is the degree to which an individual substance or a mixture of substances adversely affects living organisms. Adverse effects occur after short-term exposure (acute) or over an extended period (chronic), and may be lethal or sub-lethal (e.g. inhibition of growth or reproduction). One common method of measuring the toxicity of a particular chemical substance is to expose a relevant test organism to a concentration gradient of that particular substance. As a result, a concentration–effect relationship is established characterising the response of the organism over a wide range of toxicant doses, thus, providing information for ecological risk assessment.

Toxicity testing involves the test organism, the endpoint and the medium (Environment Canada 1999). Most standardised protocols are at the organism level (e.g. bacteria and algae, vascular plants, invertebrates or higher organisms). However, tests may also be carried out at other levels of biological organisation, such as the sub-organism or community level.

Test endpoints are a measure of the presence or absence of toxicity and they are related to a measurable characteristic of the test organism, which is affected by toxicants (e.g. percentage of growth inhibition or level of mortality). It must be noted that the measured parameters have to enable the reliable determination of the test endpoint. For example, fluorescence measurement of chlorophyll A is used to determine the biomass production (growth) of algae. However, the chlorophyll A content of cells is subject to variation due to changing environmental conditions. Therefore, it is not clearly related to growth and the use of other methods to determine algal biomass (e.g. counting of cells) may be more appropriate. Error already introduced by measuring inappropriate characteristics when determining the test endpoint will be propagated in the subsequent phases of the assessment process (extrapolation from measurement to the assessment endpoint).

It is important to distinguish between assessment and measurement endpoints (Gaudet 1994). An assessment endpoint is an environmental value that has to be protected. If the risk assessment process results in an unacceptable risk for the defined environmental value, then risk reduction measures (e.g. remediation of site) are required. A measurement endpoint is a measurable environmental characteristic, such as the quantitative summary of the results of a toxicity test or a biological survey (Suter 1993). If assessment and measurement endpoints are not the same, it is necessary to constitute a quantitative relationship between these to enable the extrapolation of measured effects to the threatened environmental characteristic.

Usually a selected test organism has a requirement for a specific environment. Nevertheless, adaptation of tests for the evaluation of samples that do not represent natural environmental conditions of the test organism is possible. For example, terrestrial organisms may be exposed to aqueous samples (e.g. plants in hydroponics), and soil solutions or elutriates from solid samples are often used in aquatic tests (e.g. freshwater algae, luminescent bacteria). However, drawing conclusions of ecological relevance from tests with organisms, which are not naturally occurring in the test medium, is questionable.

10.2.1   Level of organisation

Endpoints for environmental assessments may be selected at different levels of biological organisation:

- Sub-organism.
- Organism.
- Population.

- Community.
- Ecosystem.

At the sub-organism level, biological measures have a high diagnostic potential and are sensitive to contaminants. They have the fastest response time following exposure to a stressor and exhibit a high early warning potential for adverse environmental effects (Gaudet 1994). However, sub-organism level responses can rarely or only with difficulty be related to effects at high organisational levels, making it difficult to predict effects that may occur in the field. Therefore, such tests may be of poor relevance to the assessment endpoint. In contrast, tests at the community or ecosystem level are supposed to provide information that is more relevant to the complex processes occurring at contaminated sites including interaction with other populations (e.g. predators) and to account also for indirect effects such as reduction of available food due to toxicity. Changes in structural and functional properties can only be observed at the ecosystem level. Ecosystem-level tests may be performed at the laboratory-scale using microcosms that represent model ecosystems. However, according to Suter (1993) the applicability of ecosystem-level tests may be limited as the relationship between effects observed in the test and the responses of ecosystems exposed in the field is poorly understood. Field conditions can hardly be established in the laboratory and the level of standardisation of methods is poor. Ecosystem-level tests are usually more time consuming and expensive than (sub-)organism tests. Moreover, it has to be considered that measures at higher levels of organisation may not be diagnostic of contaminants due to the adaptation and compensation reactions of communities and ecosystems. Single organisms may suffer severe damage even before a change in functional properties of the ecosystem (e.g. respiratory, biomass production) is measurable (Suter 1993; Environment Canada 1999).

The complexity of the test system is reduced when specific organisms instead of organisms at higher levels of organisation are used. Toxicity tests using single species provide results, e.g. on survival, growth or number of off-spring, that are relevant for the estimation of field effects. The detection of responses is facilitated and the interpretation of results is straightforward when compared to more complex systems. Standardised test protocols are readily available and may be employed with reasonable expenditure and effort, and offer sufficient precision. The organism-level approach is considered to be a sound compromise in terms of sensitivity, ecological relevance and predictive potential, and is therefore suggested for use as a measurement endpoint, whereas assessment endpoints are usually defined at the population or higher level (Suter 1993; Gaudet 1994).

10.2.2 Biological responses

In general, selected tests should include responses that allow a prediction or extrapolation of effects to an assessment endpoint, which is usually at a higher level of biological organisation. Thus, responses that are meaningful for higher levels of organisation should be chosen. In ecological toxicity testing, acute and chronic responses are predominantly related to exposure duration. Acute effects are measured after a short period of toxicant exposure relative to the lifetime of the test organism. Chronic toxicity refers to effects measured after an extended exposure time that usually accounts for at least 10% of the organism's life span. As life spans vary considerably for different organisms, acute or chronic exposure cannot be linked to a generically applicable time period. Acute plant tests may last for one week, a period that represents various lengths of life spans for other photosynthetic organisms such as algae, but does not cover a relevant proportion of the life cycle of a tree. Generally, with decrease in size of organism a decrease in life span is evident. Thus, the establishment of life cycle tests with exposure over the whole life period is possible within a reasonable time frame when using small test organisms. A further possibility to reduce test duration is to focus only on the measurement of sub-lethal effects at the sensitive developmental life stages. The measurement of mortality is common to acute toxicity tests, whereas sub-lethal effects are usually measured with chronic tests. However, it has to be emphasised that *acute* and *lethal* should not be used as synonymous terms, with the same being true for *chronic* and *sub-lethal*. Most sub-lethal effects may also be measured after acute exposure if sensitivity is sufficient.

If possible, chronic toxicity tests should be applied since they usually represent naturally occurring exposure conditions in a more appropriate way. Moreover, chemicals that do not exhibit acute toxicity may cause toxic effects after chronic exposure. Using chronic tests would prevent the need to extrapolate acute test data to chronic values thus reducing uncertainty of assessments. There are few situations where acute test data may be appropriate to sufficiently describe field exposure. Examples are the single release of a toxic effluent to a river with rapid disappearance of pollutants or the temporary presence of target organisms at a polluted habitat (e.g. migratory birds). Recent efforts in toxicity testing have focused on the development of short-term tests to enhance their application in routine testing of contaminated soils. The use of test organisms with short life spans also enables the accomplishment of chronic tests within short periods.

Physiological effects. In ecotoxicology, physiological effects are usually measured at the sub-organism or sub-cellular level. They include changes in distinct enzyme activities (e.g. phosphatases, dehydrogenases) and alterations of rates of multi-enzyme reactions (e.g. respiration, photosynthesis), as well as adenylate energy charge that reveals information on the energy available for maintenance and growth processes. In general, enzymes are key elements in biochemical reactions, and their inhibition may result in severe metabolic dysfunctions and cell death. However, interpretation of results may be impeded as particular enzyme activities are considerably influenced by test conditions and test preparation, and enzymes can still remain active after the death of organisms (Römbke & Moltmann 1996). It has to be emphasised that it may be difficult to relate physiological responses of pollutant exposure to assessment endpoints, in particular to effects at higher levels of biological organisation (e.g. communities).

Survival. The extent of mortality amongst test organisms due to toxicant exposure is used frequently to describe the level of effect. Mortality may be measured after acute or chronic exposure. However, mortality is commonly associated with acute lethality after short-term toxicity testing. If a concentration series of the toxicant is tested, results are usually expressed as LC_{50}, the concentration that is lethal to 50% of tested organisms (Gaudet 1994). Mortality may be difficult to verify, particularly in case of small organisms. For example, in the acute *Daphnia* test, immobility is used as test endpoint instead of mortality.

Growth. Production of biomass is a fundamental characteristic of ecosystem development and is, therefore, frequently used in laboratory toxicity tests. Usually the growth of test organisms is directly reduced by toxicants. However, secondary effects such as inhibition of nitrogen fixation may also reduce plant growth due to nutrient insufficiency. If necessary, secondary effects should be considered. However, laboratory toxicity tests using individual test species focus mainly on direct effects and therefore, results should be used with care if predicting indirect effects in the field, and verification of estimated hazards using further lines of evidence is strongly recommended (Suter 1993).

Reproduction. Impact on reproduction is assessed by measuring the change in the number of offspring, thus having major consequences on the fate of the respective population. Measurement of reproduction can be performed in various ways including counting eggs or cocoons, determination of hatching rate or counting juveniles. Reproduction endpoints reflect field conditions appropriately, in particular for soil organisms since their reproduction takes place at the contaminated site. However, reproductive success also includes the survival of neonates and juveniles in this contaminated environment. These early life stages are regarded as very sensitive and should therefore be considered in toxicity testing (Environment Canada 1999).

10.2.2.1 Statistical estimation of endpoints

In general, two different types of statistical measurement endpoints exist (Suter 1993). The first type defines a level of effect, the second one is based on hypothesis testing. In the effect-level approach, the relationship of exposure (e.g. pollutant concentration, duration) to response (e.g. survival, growth) is determined by fitting a regression model equation to the set of test data and interpolating to the effect level of interest (e.g. LC_{50}, EC_{50}) (Moore & Caux 1997). LC_{50} is defined as the median concentration of a chemical in a defined medium (water, soil, sediment) that is estimated to be lethal to 50% of the tested organisms. The EC_{50} (median effective concentration) refers to the concentration that results in the occurrence of sub-lethal effects for 50% of the test organisms. In addition, exposure duration has to be specified and indicated when reporting effect levels (e.g. 24 hours LC_{50}). It has to be considered that LC and EC levels are related to effects (Keddy *et al.* 1994; Environment Canada 1999). Each individual organism has to be classified as either showing that effect or not (e.g. survival, reproductive failure, avoidance reactions). A reduction in an effect such as growth or light emission is not a quantal but a quantitative effect. Such effects are represented by IC_x levels (inhibiting concentration) that estimate the concentration at which a specified percentage (e.g. 20 or 50%) of impairment occurs. For example, the algae growth-inhibition test requires the calculation of an IC value (e.g. IC_{50}) because it focuses on the change of total number of algal cells (inhibition of biomass production) in the test solution and not on the reproduction of individual cells.

The second category of statistical measurement endpoints compares the effects at particular exposure concentrations to a non-exposed control. A null hypothesis is adopted, that there is no significant difference between effects measured for toxicant-containing samples and the control. Usually an α of 0.05 is selected defining the probability of rejecting the null hypothesis if it is true. The lowest concentration that shows a significant effect is defined as the lowest observed effect concentration (LOEC). The no observed effect concentration (NOEC) is the immediately lower concentration used in the test set-up. Effects measured at this concentration do not show a significant difference when compared with the control. NOEC and LOEC have been used in ecological risk assessment either directly or to derive further values. For example, Environment Canada (1999) defined a threshold observed effect concentration (TOEC) as the geometric mean of the NOEC and LOEC. TOEC is an attempt to estimate the concentration at which the toxic effects start.

The use of NOECs and LOECs has been questioned as they have some major limitations. Their values depend very much on the concentrations that have been selected in the test set-up. Large concentration steps will result in a large difference between NOEC and LOEC. Moreover, poor test design (e.g. low number of replicates) will result in increased variance of effects and the acceptance of the null hypothesis. As a consequence, the toxicity of a chemical may be underestimated. Statistical measurement endpoints obtained from ecotoxicity testing may be used to derive predicted no effect concentration (PNEC) levels that are employed in ecological risk assessment for chemicals.

10.2.3   Ecotoxicity tests and bioassays

Ecotoxicity tests have been developed to characterise the toxicity of individual chemicals. Data of these single chemical tests have been used to derive quality criteria (chemical concentrations). These criteria have been used for the evaluation of the non-harmful release of xenobiotics or the assessment of contaminated media by comparing them to modelled or measured concentrations.

Ecotoxicity tests have been adopted for the toxicity evaluation of environmental samples and are referred to as contaminated media tests (Suter *et al.* 2000) or bioassays (DECHEMA 1995; Ferguson *et al.* 1998). The more general term *toxicity test* is also applied to single species tests but it does not distinguish between tests for chemicals or environmental samples. In this chapter, the term *bioassay* is used for toxicity tests that are applied to environmental samples with no or insufficient information on type and concentration of pollution. In contrary to bioassays, the term *ecotoxicity test* is used in this work to describe tests that focus on the toxicity evaluation of individual chemicals. For ecotoxicity tests, the chemicals of concern are usually added to a standard test medium and a concentration–effect relationship is determined with the focus on prediction of the environmental effects. In many cases, test protocols for bioassays have been adopted from protocols for ecotoxicity testing. However, test data may not be expressed in the same manner since it is often not feasible (and not necessary) to calculate toxicity levels or values (e.g. EC_{50}, NOEC) for bioassays. Relating effects of bioassays to pollutant concentrations would require complete chemical analysis of environmental samples as well as understanding and quantification of pollutant interactions.

10.3   The use of bioassays for soil evaluation

Many countries have established lists with priority pollutants (Ferguson & Kasamas 1999; ATSDR 2001) focusing on substances that are hazardous to human health and/or the environment. These lists refer to chemicals with known toxicity at least for particular receptors. If information on the nature of pollution is available, chemical analysis may focus on selected chemicals. In case of waste sites or if data on site history are scarce, comprehensive chemical analysis would be required to reduce the risk of disregarding substantial pollutants. This is expensive. The application of bioassays may circumvent extended routine analysis by indicating toxic samples directly.

When discussing bioassays, focus is given to laboratory tests applying organisms to potentially polluted soil that is sampled at the suspected site and transferred to the laboratory (*ex situ* tests). Usually cultivated test organisms are added to the contaminated media. This is a widely used practice and numerous standard protocols are available. There exist further possibilities to investigate toxicity of contaminated media, such as planting test organisms directly at the site (*in situ* test). Another way is to sample organisms at the contaminated site and analyse them in the laboratory for bioaccumulation, or abnormalites etc. (Suter *et al.* 2000). A special situation for performing bioassays is to bring soil samples with naturally occurring organisms to the laboratory and analyse effects to the indigenous biota in defined laboratory conditions. In this case, effects often represent ecosystem functions such as soil respiration or nitrification capacity. However, it has to be considered that this approach is very susceptible to inaccuracy due to inappropriate sampling methods, storage conditions and sample preparation. Moreover, organisms that are exposed to toxicants for long periods may adapt and develop resistance to the exposure conditions. In particular, when measuring community or ecosystem functions, compensation mechanisms may mask toxic effects. In fact, this phenomenon is used to detect earlier contaminant exposure to microbial communities by measuring pollution induced community tolerance (PICT) (Rutgers & Breure 1999).

Advantages of bioassays

- Direct measurement of lethal or sub-lethal effects.
- Results may be immediately understandable and convincing.
- Applicable if toxicant is not analysed, not known or lacks toxicological data.
- Less costly than chemical analysis for complex pollution.
- Measurement of combined toxicity (interactions for mixtures).
- Consideration of chemical form of contaminants.
- Implicit consideration of bioavailability of organics and metals.
- Site specific assessment.

Limitations of bioassays

- No identification of toxic substances.
- Extended test duration compared to simple chemical analysis.
- Limited number of species tested.
- Extrapolation to field conditions (e.g. multiple stress).
- Specific sampling, storage and preparation requirements.
- Increased costs for comprehensive testing.
- Apparent toxicity due to confounding factors and unsuitable reference soil.

Advantages. An outstanding benefit of bioassays is that effects are determined directly rather than concentrations (which again are used to predict effects). The test organism is exposed to the pollutants, and toxicity is indicated if the sample is hazardous to the test organism. Toxic effects are more easily understandable than concentrations of analysed pollutants. For example, if 100% of the tested earthworms die in the soil sample, the bioassay demonstrates rather clearly that toxicants are present. A list of pollutant concentrations would reveal hazardous potential only after the values have been compared with, for example, benchmark concentrations (which may again be based on toxicity testing). Even if the toxicant is unknown or was not detected using chemical analysis, the test organisms will indicate toxicity. Hazards may also be determined for samples containing substances with no available toxicological data.

Detailed investigation of soil samples with comprehensive analysis is possible but expensive. In particular, if multiple pollutants are present, bioassays may be less costly (Environment Canada 1999).

Furthermore, bioassays are of benefit for the assessment of complex mixtures. Usually, more than one individual pollutant occurs at a contaminated site and besides additive, synergistic (more than additive) or antagonistic effects (less than additive) may also determine the overall toxicity. Interactions among pollutants are complex, and simply summing the effects of identified substances may be misleading for the assessment of the actual hazard. In addition, the chemical state (e.g. association with organic or inorganic fraction) as well as the ionic charge (e.g. oxidation state of metals) of the pollutant that determines the intrinsic toxicity of the substance is considered. The total pollutant concentration, however, remains unchanged. While chemical methods used in routine analysis usually determine only total concentrations, the test organism is exposed to the substances as they occur in the sample. Moreover, sequestration of the pollutant may lead to a decreased toxicity due to a lower availability of the toxicant to ecological receptors. Whereas the bioavailability of metals in soil may be estimated using empirical extraction (leaching) procedures, such as the ammonium chloride technique (Wenzel & Blum 1999), the availability evaluation of organic contaminants is more complex. The bioavailability of organic pollutants in soil is influenced by the following factors:

- Physical-chemical characteristics of the contaminant ($\log K_{ow}$).
- Soil composition (amount and type of organic matter, particle size distribution and moisture content).
- Concentration of contaminant and co-pollutants.
- Residence time in soil (ageing).
- Receptor organism.

If hydrophobic organic pollutants such as high molecular weight PAH are present in soil, they are usually sorbed by soil organic matter, a polymeric material with expanded (weak sorption) and condensed (strong sorption) domains (Fig. 10.2). Moreover, pollutants may diffuse into nanopores of the soil matrix, therefore being rendered unavailable to microorganisms or other biota. These time-dependent processes are summarised as ageing, and the resulting decrease in bioavailability has to be considered when soils are spiked with contaminants for toxicity measurements. Moreover, the concentration of pollutants may influence the ageing process, and co-contaminants of low toxicity may increase the mobility of usually poorly available but highly toxic compounds.

The more available fractions comprise dissolved contaminants and the part sorbed by expanded (soft and flexible) organic matter (Luthy *et al.* 1997). Substantial research has been accomplished over the last years and is still on-going to quantify the available or non-available part of pollution (Cuypers *et al.* 2000; Loibner *et al.* 2000; Liste & Alexander 2002). However, no routine

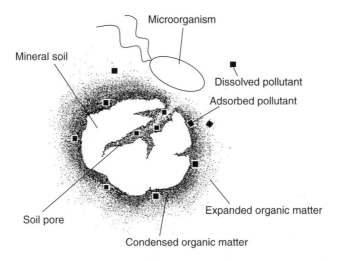

Figure 10.2 Schematic presentation of a soil particle indicating sites of pollutant sequestration.

methods have been developed that would enable the measurement of the available portion of organic contaminants. It has to be emphasised that there exists no overall measure of bioavailability as it changes with the receptor. For example, chemicals not available to plants may be available to earthworms (ingestion of contaminated soil). Therefore, chemical or physical methods that mimic bio-availability should always identify the organism or group of organisms to which the results are related. In contrast, toxic effects observed in bioassays result from the pollutant fraction that is bioavailable to that particular organism.

Bioassays applied to contaminated soil samples provide an overall toxicity measure that accounts for numerous site-specific variations. Factors that may lead to an enhanced or reduced toxicity are no longer a matter of estimation but are measured directly with the test system.

Limitations. It has to be emphasised that bioassays do not identify the substances that cause toxicity. If remedial actions follow, identification of the contaminants might be necessary in order to choose the right remediation technique (e.g. bioremediation for biodegradable compounds, pollutant extraction for non-degradables). In the case of mixed pollution, toxicity identification evaluation (TIE) may be used to identify from which contaminants or pollutant fraction toxicity is originating. In general, TIE is a technique that performs toxicity measurements for different fractions of the sample (extracts, fractionated leachates), and toxicity identification for a particular fraction allows one to form a conclusion on the presence of a particular pollutant group. However, fractiona-tion and individual testing of fractions would result in increased experimental effort and so TIE is not applied in routine testing with high sample numbers.

Bioassays are not always preferable, especially, if simple chemical analysis is sufficient (e.g. spill with known pollution) to answer questions such as what is the spatial extent of the spill. Many bioassays have an extended test duration and may only be competitive if they can substitute for comprehensive chemical analysis. Nevertheless, one should keep in mind that bioassays provide additional information not achievable with conventional chemical analysis, which could be a compensation for longer test duration.

Only a limited number of organisms can be tested in the laboratory and they may have to serve as surrogates for key species. Test conditions in the laboratory are standardised and usually differ from those present at the contaminated site. Fluctuations in temperature, moisture, available food as well as predatory activities cause additional stress and may result in altered resistance of organisms.

A general problem of bioassays is the occurrence of confounding factors such as pH or salinity that may interfere with the measurement parameter. For example, growth inhibition of algae may be a result of increased pH and not due to the presence of pollutants. It must also be noted that the selection of an appropriate reference soil is of utmost importance for the accuracy of the test results since the measured effect is usually related to the reference (Environment Canada 1999). Any enhancing or inhibiting effect that occurs in the reference but not in the tested soil sample will lead to a falsified test result. Finally, it should be mentioned that specific sampling, storage and pre-treatment of samples might be required for bioassays, in particular if microbial soil functions are the subject of the testing (Becker & Ginn 1990).

10.3.1   Ecological assessment and toxicity screening

Bioassays may be employed for ecological assessments or toxicity screening. If bioassays are used in the framework of an ecological risk assessment, they have to be predictive for the assessment endpoint. Therefore, ecologically relevant criteria govern the selection of the tests. The prediction of field effects usually at higher levels of organisation requires a comprehensive test set-up and should also include chronic exposure to contaminated soil. Bioassays with increased ecological relevance are usually performed at higher tiers of the assessment process.

With the aim of using bioassays in routine analysis with a high sample throughput, test protocols have been improved or newly developed for a rapid and reliable identification of potentially contaminated samples (Persoone et al. 2000). Toxicity screening does not necessarily require an ecological assessment endpoint as minor or no priority is given to ecological considerations. Screening tests may provide information as to whether toxicants are present or not, rather than what ecological consequences may be caused by pollutants present.

The distinction between the use of bioassays for toxicity screening and ecological assessments is not always clear and feasible. With the simplicity, sensitivity and reliability that many screening tests offer, it is tempting to use them for the evaluation of complex ecological processes. However, it has to be noted that many screening tests are applied using aquatic organisms, hence, the significance of results for soil-related assessment endpoints is usually vague.

10.3.2   Selection criteria for bioassays

The relative importance of selection criteria for screening tests and ecologically predictive tests is indicated in Table 10.1. Ambiguous items are discussed below.

A high level of sensitivity to a wide range of toxicants is required for screening tests in order to minimise false negative results. Since one sensitive species for all possible contaminants and scenarios does not exist, the application of more than one species (test battery comprising several sensitive bioassays) is recommended in order to cover a wide range of pollutants.

It has to be noted, that the establishment of a test battery of ecologically relevant tests addresses particular needs such as the consideration of key species that are representative of particular soil communities or soil functions that have been defined in the problem definition phase of the risk assessment procedure. Another approach is to define a simplified hypothetical ecosystem that includes organisms from different trophic levels (e.g. producers, consumers, decomposers) and/or phylogenetic groups (e.g. microorganisms, plants, invertebrates). High sensitivity is not that important for ecologically relevant tests as long as sensitivity is sufficient to enable extrapolation to the assessment endpoint. False negative results (pollutants present but no toxicity observed) are not problematic if the receptor to be protected is not harmed by the pollution (e.g. measurement and assessment endpoint are identical or exhibit comparable sensitivity).

The availability of sufficient and suitable test organisms is a major criterion for the practical application of a specific test. If not commercially available (e.g. various test kits including test organisms that can be cultured upon demand

Table 10.1 Importance of selection criteria for distinct test applications

	Toxicity screening	Ecological assessment
Ecological relevance	Not important	Important
High sensitivity	Required	Dependent on assessment endpoint
False negative responses	Problematic	Problematic
Availability of test organisms	Important	Important
Availability of reference soils	Important	Important
Standardised protocols	Important	Helpful
Short test duration	Important	Preferable
Ease of handling	Important	Preferable
Low costs	Important	Preferable
Robustness	Important	Preferable
Reproducibility	Important	Important
Reliability	Important	Important
Repeatability	Important	Important

are available from MicroBioTests Inc., Nazareth, Belgium; see also Bioassays), organisms have to be cultured in the laboratory requiring trained personnel and laboratory space with appropriate incubators. To avoid toxicant adaptation of test organisms, culturing should be accomplished in a separate facility or isolated compartment of the laboratory, not used for the manipulation of contaminated samples. Maintenance of cultures should not be demanding, and test organisms should exhibit a high reproduction rate and short life cycles with a tolerance for changing environmental conditions. Organisms should not be of public interest (e.g. no endangered species) and methods for the control of the sensitivity (e.g. the use of reference toxicants in test applications) should be available. The selection of an appropriate reference soil is crucial for most bioassays. The optimum reference soil differs from the contaminated soil only by the absence of pollution. Growth conditions and survival of organisms may be notably influenced by the soil environment. For example, plant growth varies considerably with soil characteristics and selecting an unsuitable reference soil may result in false negative or false positive data. Reference soils are sampled at a non-polluted location at or close to the contaminated site, whereas soils selected from a collection of well-described model soils are usually named control soils. A limited number of natural standard soils is commercially available (LUFA, Speyer, Germany, www.lufa-speyer.de). At least pH, soil organic matter content, soil texture and nutrient condition as a minimum should be comparable for reference (control) and test soils. Specific tests may require matching of further parameters. Elutriate tests should always be performed by comparing test data to a blank (aqueous medium). The resulting apparent toxicity, originating, for example, from colouration of the test elutriate (interference with spectroscopic methods) may be avoided by relating test data to elutriates from an appropriate reference soil. This will also account for the masking of toxic effects caused by, for example, the presence of nutrients in the test elutriate resulting in enhanced growth of algae.

The influence of the reference soil on the test result depends on the type of effects measured. Generalisation may be difficult but usually severe effects (lethality) are less influenced by the soil matrix than sub-lethal effects (e.g. mortality of earthworm compared to growth of plants).

10.3.3 Scope of application

Bioassays have been widely used for the evaluation of potentially contaminated sites – with varying degrees of success. For test applications, it is crucial to determine in advance what information is expected from the test and how the test data have to be processed in order to obtain this information. The percentage of plant growth reduction and bioluminescence inhibition of two bioassays, for example, will not be conclusive for remediation needs but may indicate the presence of pollution and the necessity for further investigation.

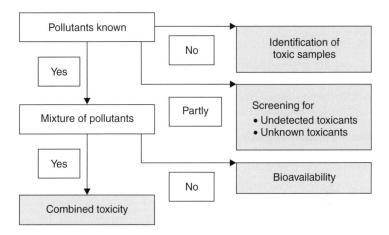

Figure 10.3 Application tree for bioassays. Grey boxes indicate information that can be derived from the use of bioassays.

10.3.3.1 The role of bioassays in soil testing

Depending on the type and extent of data available from historical investigations or chemical analysis, different information may be obtained from bioassay applications (Fig. 10.3).

Pollution unknown or information limited. If no data are accessible on the presence of pollution, bioassays may be used for toxicity screening in order to identify polluted samples. Moreover, if only insufficient data from chemical analysis are available, bioassays may identify samples that require further analysis (presence of undetected or unknown toxicants). Although bioassays respond to a wide range of toxicants, success in identification of contaminated samples may be reduced if false negative results are achieved (see below).

Complete information on pollution available. If all contaminants present in the sample have been analysed using conventional physical or chemical methods, the application of bioassays provides further information e.g. on toxicity interactions and bioavailability. It has to be emphasised that for multiple contamination, bioassays reveal an overall combined toxicity response to the bioavailable part of each contaminant present in the mixture. The identification of the bioavailable fraction of a particular substance by relating the measured effect to the measured concentration of the chemical is only possible for a single substance contamination (only one toxic chemical present). Although it may be tempting for risk assessors, the response of multiple pollution cannot be used to calculate the bioavailable fraction (percentage of the total concentration of a compound) of a particular chemical from the mixture, unless toxicant interactions can be excluded or quantified and the availability of co-pollutants is known and understood.

10.3.3.2   Application of bioassays for toxicity screening

Bioassays used as screening tools aim at the identification of toxic samples. Short test duration, sensitivity, reliability and ease of use are favourable characteristics for test selection. Screening test applications include:

- Determination of spatial extent of contamination.
- Screening of samples with unknown contamination.
- Assessment of mobility of contaminants.
- Estimation of bioavailability to predict bioremediation performance.
- Monitoring of detoxification measures.

One step in assessment of a contaminated site is the determination of the spatial extent and distribution of the contamination. For heterogeneous pollution, screening tests may determine hot spots and areas with no or low toxicity within a short time period at reasonable cost. However, if historical data provide sufficient information on the type of contamination, then chemical analysis may be preferable when focusing on single toxicants or simple groups of toxicants (e.g. BTEX); ecotoxicological information can be ascertained from literature or databases.

Bioassays may also be used to screen for toxic samples in order to identify those that require extended chemical analysis. Moreover, screening tests can provide information as to whether leaching of toxicants to the groundwater will occur. The leaching potential can be determined by the preparation of aqueous soil extracts. Soluble toxicants, if not retained by the soil, will elute and cause a toxic response. Various short-term aquatic tests are available that can be applied to aqueous extracts (elutriates and leachates). Toxicity of elutriates indicates a potential threat to groundwater and/or surface water. In addition, the leaching characteristics of polluted soils may also provide information about the feasibility of bioremediation techniques. For soil systems, leachable contaminants are considered to be bioavailable, since pollutants present in elutriates are accessible to microorganisms involved in the degradation process. Bioremediation will only be effective, if pollutants are bioavailable. However, it has to be emphasised that immobile (non-leachable) contaminants may also be available to microorganisms, thus, concentrations in elutriates may underestimate bioavailability. Nevertheless, toxicity measured in elutriates may be a good indicator for the bioremediation potential of biodegradable soil contaminants unless inhibiting conditions are present.

Remediation aims at the reduction of risk. The disappearance of native pollutants during remediation does not necessarily result in a reduction of risk. Screening tests may be valuable monitoring tools to evaluate remediation measures as they may indicate, for example, an undesirable formation of toxic intermediates or an unwanted increase of pollutant mobility.

False negative results. The detection of no effects for toxic samples may have the following reasons:

- Lack of sensitivity.
- Inappropriate control (constituents in sample compensate adverse effects).
- Inappropriate test conditions (volatilisation, degradation of contaminants).
- Antagonistic effects.

Insufficient sensitivity is the most relevant factor, others are considered to reduce toxicity responses but are hardly able to totally mask toxicity. Estimation of the probability of obtaining false negative results will help to decide whether further investigation is required.

Considering surface soil at old contaminated sites, volatilisation and degradation will have already occurred in the field. Volatilisation and degradation are usually not issues for screening tests; however, if there is a possibility that these processes confound results, then they should be considered in the test set-up, in particular for long-term tests (duration of several weeks).

10.3.3.3   Application of bioassays for ecological assessments
The evaluation of threats posed to ecological receptors or functions is probably the most significant reason for the application of biological test methods. There exist different approaches for using bioassays to perform risk assessments and to draw conclusions for risk management decisions.

Integration of bioassays in risk assessment frameworks. If the application of bioassays is included in a framework for the estimation of the probability of harm to ecological receptors, tests then have to be selected that enable extrapolation from the measured test endpoint to the assessment endpoint as defined in the problem formulation. Modelling of exposure is not necessary as it is measured directly by the use of bioassays. There is no need to make assumptions about bioavailability; it is considered individually for each sample. Effects are measured and do not have to be predicted from measured pollutant concentrations. The measured effect is the summary of all effects of known and unknown pollutants (including different chemical forms) and their interactions (combined toxicity). As exposure and effects are measured directly with site samples, uncertainty, usually high in the analysis phase (exposure and effect assessment) may be reduced remarkably. In the risk characterisation step, the probability that the effects measured in the laboratory will occur in the field has to be quantified. For example, based on bioassay data generated using *Lumbricus terrestris*, the probability that the earthworm population at the site will be reduced by 30% is estimated to be 80%. This still does not answer the question whether remediation is required or not, unless a trigger value for remediation measures (e.g. a 30% reduction of the population in the field that occurs with a probability greater than 95%) has already been defined in the problem formulation phase. It has to be mentioned that the level of acceptable risk is a political decision rather than a scientific one.

The extrapolation of test data includes uncertainty, in particular, if measurement and assessment endpoints are not identical. However, this source of variation is even more pronounced when using chemical concentrations for the prediction of field effects. It has to be emphasised that the use of bioassays represents one line of evidence, which together with chemical analysis and/or biological field data should be used for a reliable ecological risk assessment.

Usefulness of correlating chemical concentrations to effects measured with bioassays. Effects observed for contaminated samples are usually compared with data obtained for control or reference soils, and significant differences are estimated using statistical hypothesis testing. Results may be expressed as, for example, the sample exhibits 20% inhibition compared to the control. Relating effects to concentrations of pollutants present in the sample as measured by chemical analysis may not always be possible and reasonable. However, ecotoxicologists and risk assessors are used to dealing with chemical concentrations and therefore, tend to relate test responses to concentrations. Attributing effects to concentrations of individual chemicals is difficult for samples with contaminant mixtures (which actually is the case for most sites) if combined toxicity occurs. Moreover, if comprehensive chemical analysis revealed only one pollutant being present (one single species of a chemical substance), there is still some likelihood of not having detected additional toxic compounds that may cause combined toxic effects. Risk assessors are probably used to quantifying such risks, however, if uncertainty can be reduced, this should be done.

Moreover, the testing of different concentration levels would be required for the calculation of effect levels for a single known chemical. It may be difficult to find concentration gradients at the contaminated site comprising the same type of pollution in an equivalent matrix. Diluting collected samples with a non-contaminated control or reference soil is another possibility but requires an appropriate non-contaminated soil. In general, testing of concentration series multiplies the number of samples to be tested by the number of dilutions. Keeping also in mind the replicates necessary for statistical data evaluation, the resulting high number of tests might be an obstacle for the application of bioassays in contaminated land assessment at least in preliminary evaluations with high sample numbers.

Calculating effect levels for polluted soils rather than for chemical concentrations is another option and would generate results such as the LC_{50} for earthworms in the tested contaminated soil/reference soil mixture amounts to 30% of the contaminated soil and 70% reference soil. However, this approach also requires dilution with a reference soil and it may be questioned for its practical relevance. Effect levels may be useful for the evaluation of mobile media such as effluents that are subject to dilution in the environment. However, soil at contaminated sites is usually static compared to aquatic media. There may be some surface movement of contaminated soil caused by erosion or flooding, etc., but in the sub-surface frequently occurring dilution processes of contaminated soil do not exist unless anthropogenic activities such as excavation take place. Considering the problems related to the establishment of appropriate dilution series for contaminated soil and taking into account that soil dilution naturally occurs only to a minor extent, there seems to be no urgent need for establishing concentration-effect relationships and for calculating effect levels when using bioassays, except that the legal framework considers mixing of polluted and non-contaminated soil as a suitable remediation measure. However, this may not be the case in many European countries.

Test batteries for toxicity evaluation. A common approach for the evaluation of a particular soil function is the definition of the ecological structure that is responsible for this function. The establishment of a test battery that includes keystone species may be used to predict the impairment of the soil function. However, species interactions are not considered when applying test batteries. Nevertheless, this very pragmatic approach seems to be a reasonable compromise between ecological and practical needs. DECHEMA (1995) defines soil functions ('pollutant retention' and 'habitat for plant growth and biocenoses') as well as five different soil uses ranging from soil under sealed areas to agricultural land. Test batteries for the evaluation of each soil function are suggested. Results of each test are summarised and compared to a score, that is considered to be protective for the particular soil function (habitat) for a selected soil use (e.g. landfill covering or green area). The probability of the occurrence of effects is not addressed in this approach. There is also a more recent issue (German language version only) available from DECHEMA (2001). Major modifications relate to the inclusion of a set of genotoxicity tests, the refinement of test batteries and the comparison of test data to a standard soil (e.g. LUFA soil 2.2) rather than to predefined values (e.g. mg CO_2 production in the respiration test).

Soil function tests. Soil is often valued for its function rather than the structure of inhabiting populations or communities. Functional properties of ecological importance comprise e.g. carbon and nutrient cycling, the support for plant growth and the maintenance of biodiversity (Suter *et al.* 2000). Predicting soil quality based on soil functions may be difficult if only single species tests are used. Therefore, multispecies tests usually applied in microcosm or mesocosm studies are suggested. Moreover, the use of microbial community tests may reveal information on the nutrient cycling capacity of contaminated soils.

10.3.4 Test media: solid samples and aqueous extracts

Soil ecology generally addresses biota at the surface of soil (e.g. plants) or the top soil layer (about one to two metres). Soil is not abiotic below this depth but diversity and abundance of (macro)organisms decrease with depth. For many sites (point pollution rather than diffuse pollution) major contamination occurs below surface and usually goes down to the aquifer, potentially spreading with ground-water. The question of whether these soil zones should also be subject to eco-logical risk assessment is still a matter of discussion: does it make sense to test deep soils using earthworms or plants? If excavation and surface deposition (as part of the remediation strategy) is not intended, toxicity screening may be performed revealing valuable information that is useful for other than ecological purposes.

Whole soil. A growing number of standardised methods are now available for terrestrial organisms using whole soil as test medium (ISO/DIS 15799 2001). Soil consists of all three primary compartments comprising solids, water and air, with each of them being a possible pathway for pollutant transfer to the living organism. Uptake of pollutants may occur via dermal contact, by ingestion of whole soil or soil components (e.g. soil pore water, soil organic matter), or by inhalation of volatile contaminants. Although solids and air may account for exposure of particular organisms and pollutants, the most frequent uptake of contaminants may occur from soil solution.

Soil pore water. Tests using extracted soil pore water may give some insight into naturally occurring exposure routes. However, contaminants present in the pore water do not always reflect the total available amount of pollutants. Since the equilibrium concentration of hydrophobic contaminants is usually very low in soil pore water, uptake of contaminants from the water phase by the receptor would cause further pollutant desorption from the soil matrix resulting in successive uptake. This time dependent pollutant release should be considered when performing bioassays. Moreover, sampling of soil solution may be laborious in particular if large volumes of pore water are needed for the test.

Aqueous extracts. Various bioassay protocols are available for testing soil elutriates or leachates. Historically these have previously been available for aquatic tests. However, it may be inappropriate to project effects generated with aquatic species to soil organisms or ecosystems. With the development of whole soil bioassays, the use of elutriate or leachate tests focuses now on the prediction of threats to groundwater (leaching of toxicants) or surface water (contaminant run-off). Tests using aqueous extracts address pollutants soluble in water, and therefore are a measure of mobile toxicants.

Elutriate preparation is a crucial step for achieving reliable test data. Extraction times (preferably involving end-over-end shaking) range from approximately 30 minutes to 24 hours. Soil to water (w/w) ratios range from 1:2 (preferable, DECHEMA 2001) to 1:10 (DIN 38414-4 1999). High ratios are considered to poorly reflect naturally occurring conditions and may result in a high dilution of soluble contaminants. Removal of particulate matter from elutriates is important for test reproducibility. Sedimentation is not sufficient to remove all solids. Filtration is highly problematic for hydrophobic contaminants that may be bound to the filter material. If filtration is the only option, the use of glass fibre filters is recommended (since they only give rise to low adsorption of organic pollutants). For particulate matter removal, centrifugation at 5000 g for 20 minutes should be an appropriate method to produce satisfactory results (dela Cruz *et al.* 2000). However, it has to be emphasised that only glassware should be used when hydrophobic contaminants are expected (e.g. samples from gas works containing PAH). This applies not only to centrifugation tubes but also to all lab-ware coming into contact with the elutriates.

10.3.4.1 Pre-treatment and storage of contaminated samples

Changes in toxicity may occur during pre-treatment and storage of contaminated media. This may be due to either a loss of toxicants or changes in the bioavailability. The loss of contaminants during storage may be a result of volatilisation or biodegradation. Samples suspected of containing volatiles require special procedures not only for storage and pre-treatment (e.g. drying, sieving) but also for the application of bioassays with standard test procedures often not being applicable for such samples.

Sieving is required for most soil samples prior to testing or analysis. The mesh size should refer to the intended test application but is usually 2 mm with the finer soil fraction used for the tests. Bioassays requiring large amounts of soil (e.g. plant or earthworm tests) may use larger mesh sizes e.g. 5 mm (ISO 11269-2 1995) in order to reduce the sieving effort.

Pre-treatment such as homogenisation or sieving may also result in an *activation* of degrading microorganisms by providing suitable growth conditions (e.g. oxygen supply). Rost *et al.* (2002) observed significant biodegradation of PAH (up to five rings) in soil within two weeks of storage at temperatures as low as 4°C. Freezing of soil samples (−18°C) may suppress biodegradation but will change the integrity of soil samples giving rise to an altered bioavailability. However, freezing may be used for the storage of aqueous extracts. DECHEMA (2001) suggests −18°C for elutriate holding times longer than 7 days. Addition of microbial inhibitors (sample preservation) may be appropriate for chemical analysis but most probably result in enhanced toxicity in bioassays. Generally, storage times should be as short as possible with samples being held in the dark at 4°C. Placing samples in an inert gas atmosphere (e.g. nitrogen) to prevent aerobic breakdown of pollutants is recommended for longer holding periods of up to eight weeks (Becker & Ginn 1990).

10.3.5 Bioassays relevant for contaminated soil analysis

A selection of tests that may be employed for the investigation of soil pollution is given in 10.5 Bioassays. The listing comprises frequently used bioassays. Many of the tests described have been standardised by agencies or associations

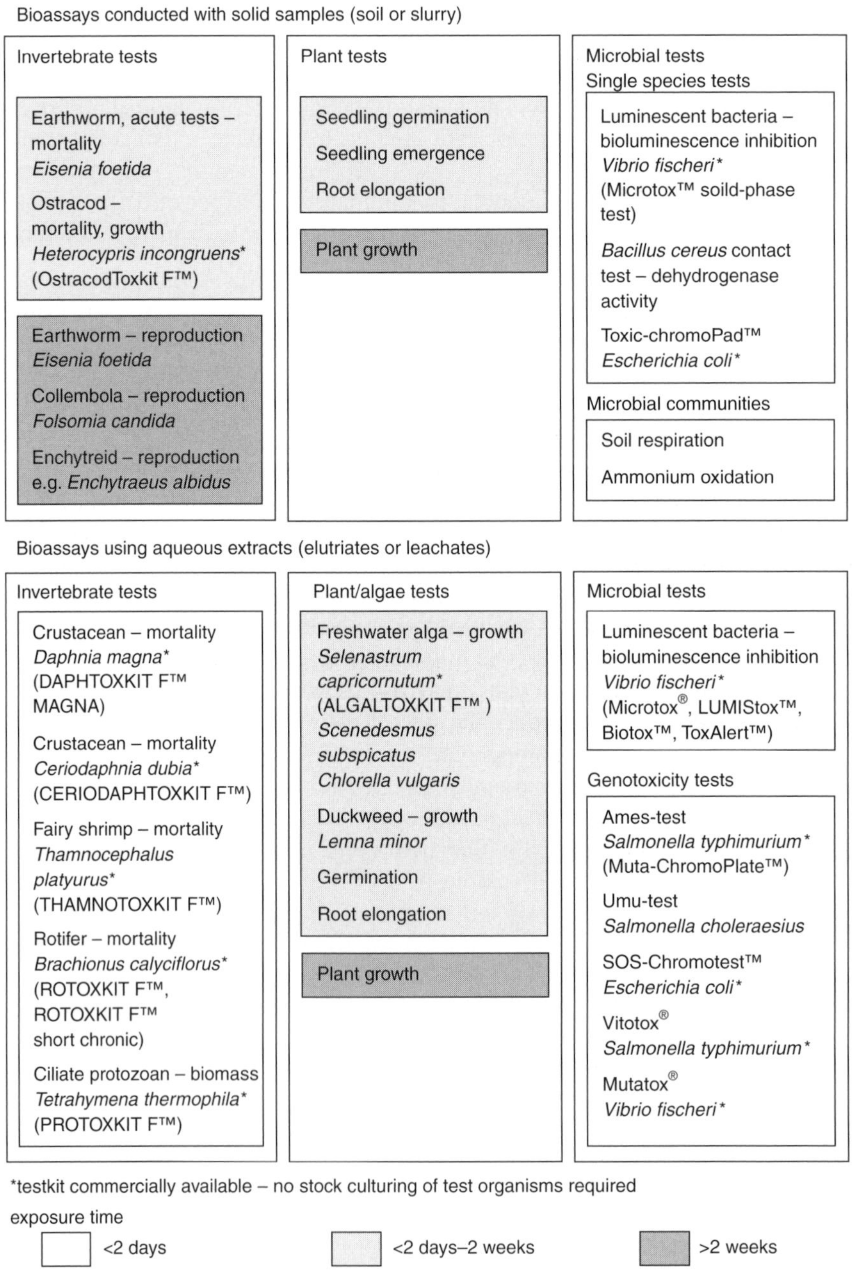

Bioassays conducted with solid samples (soil or slurry)

Invertebrate tests	Plant tests	Microbial tests Single species tests
Earthworm, acute tests – mortality *Eisenia foetida* Ostracod – mortality, growth *Heterocypris incongruens** (OstracodToxkit F™)	Seedling germination Seedling emergence Root elongation Plant growth	Luminescent bacteria – bioluminescence inhibition *Vibrio fischeri** (Microtox™ soild-phase test) *Bacillus cereus* contact test – dehydrogenase activity Toxic-chromoPad™ *Escherichia coli**
Earthworm – reproduction *Eisenia foetida* Collembola – reproduction *Folsomia candida* Enchytreid – reproduction e.g. *Enchytraeus albidus*		Microbial communities Soil respiration Ammonium oxidation

Bioassays using aqueous extracts (elutriates or leachates)

Invertebrate tests	Plant/algae tests	Microbial tests
Crustacean – mortality *Daphnia magna** (DAPHTOXKIT F™ MAGNA) Crustacean – mortality *Ceriodaphnia dubia** (CERIODAPHTOXKIT F™) Fairy shrimp – mortality *Thamnocephalus platyurus** (THAMNOTOXKIT F™) Rotifer – mortality *Brachionus calciflorus** (ROTOXKIT F™, ROTOXKIT F™ short chronic) Ciliate protozoan – biomass *Tetrahymena thermophila** (PROTOXKIT F™)	Freshwater alga – growth *Selenastrum capricornutum** (ALGALTOXKIT F™) *Scenedesmus subspicatus* *Chlorella vulgaris* Duckweed – growth *Lemna minor* Germination Root elongation Plant growth	Luminescent bacteria – bioluminescence inhibition *Vibrio fischeri** (Microtox®, LUMIStox™, Biotox™, ToxAlert™) Genotoxicity tests Ames-test *Salmonella typhimurium** (Muta-ChromoPlate™) Umu-test *Salmonella choleraesius* SOS-Chromotest™ *Escherichia coli** Vitotox® *Salmonella typhimurium** Mutatox® *Vibrio fischeri**

*testkit commercially available – no stock culturing of test organisms required

exposure time

☐ <2 days   ▨ <2 days–2 weeks   ▨ >2 weeks

Figure 10.4 Bioassay categories with respect to employed organisms and test duration.

such as ASTM, CEN, DIN, Environment Canada, ISO, OECD, OENORM and USEPA with a focus on toxicity evaluation of existing chemicals or testing of newly synthesised compounds prior to their release. However, these tests may also be used for toxicity testing of contaminated media, often without extensive adaptations of the protocol. The application of standardised test protocols usually accounts for decreased variability in data thus reducing uncertainty. However, the lack of a standardised protocol should not be a reason to exclude a test that is known to be highly relevant to a particular assessment.

Figure 10.4 represents a summary of bioassays distinguishing between whole soil and elutriate tests reflecting different assessment objectives. Whilst whole soil tests may preferably be used for ecological assessments (e.g. the investigation of habitat functions of soil), testing of elutriates provides information on the mobility of contaminants and the retention capability of soil.

The application of bioassays for toxicity screening is supported by short test periods and rapid access to sufficient test organisms. To facilitate the selection of appropriate screening tests, the duration of bioassays is indicated in Figure 10.4. Increasing grey intensities of individual boxes account for increasing exposure periods. The differentiation between invertebrate, plant (including algae) and microbial tests relates to different phylogenetic levels of test organisms that may be relevant for the set-up of a test battery. Organisms may also be selected according to different trophic levels (producer, consumer, decomposer). It has to be noted that in simplified ecosystems, usually mimicked by test batteries, overlapping of phylogenetic groups and trophic levels occurs (plants/producers, invertebrates/consumers, microorganisms/decomposers). Moreover, genotoxicity tests are listed as an own group being used commonly for the evaluation of the mutagenic potential of environmental samples with results probably more relevant for human health risk assessments.

10.4 Conclusion

The application of bioassays provides information that may not be obtained by chemical analysis. Biological tests are not intended to replace conventional soil analysis; they should rather complement each other in assisting the assessment of soil pollution. Bioassays provide a direct measure of the effects of actual exposure. There is no need to extrapolate from measured chemical concentrations to possible effects. The prediction of bioavailability and combined toxicity from chemical concentrations always results in increased uncertainty. Bioassays consider soil–pollutant and pollutant–pollutant interactions by providing an overall toxicity response. They have the potential to detect (but not identify) the presence of toxic compounds that have not been considered in pollutant priority lists so far (e.g. non-target contaminants in pollutant groups like PAH that are hazardous to ecological receptors).

The growing interest in the application of bioassays for the evaluation of contaminated land has resulted in standardised procedures for soil testing. The development of short-term bioassays and commercially available testkits promotes the application of biological test methods. However, before using bioassays, a decision on the objectives of the application has to be made. On the one hand, inexpensive and easy to use bioassays may be applied for screening purposes to identify toxic samples. On the other hand, bioassay data may be used to estimate the ecological hazard posed by polluted soil. Regulatory frameworks for the integration of biological data in risk assessments are emerging. Compared to current practices with defined generic standards for chemical concentrations, obligatory values for test endpoints (e.g. maximum allowable growth inhibition of a defined species) are not available. Such toxicological standards would support the application of bioassays in contaminated land risk assessment but require pragmatic collaboration between scientists and regulators.

Acknowledgements

The authors would like to express their gratitude to Jason Weeks (National Centre for Environmental Toxicology, WRc-NSF Ltd.) and Marion Hasinger (IFA-Tulln) for helpful comments and reviewing the manuscript.

10.5   Bioassays

Bioassays conducted with solid samples (soil or slurry)

Invertebrate tests
Earthworm – acute toxicity test

Test organism: *Eisenia foetida* or *Eisenia andrei* (red worm).
Principle/Procedure: The mortality of adult earthworms placed in potentially contaminated soil is determined.
Exposure time: 7, 14 days.
Specific equipment: temperature controlled at 20°C ± 2°C.
Test parameter(s): mortality.
Limitations: The method does not take into account a possible volatilisation or degradation of pollutants during the test.
References/Guidelines: ISO 11268-1 (1993); OECD 207 (1984).
Comments: The epigeic worm *Eisenia foetida* belongs to the composting worms. These red worms can easily be cultured in large quantities in the laboratory, thus, they are widely used and recommended in international standards. However, the ecological relevance of *Eisenia* is limited as it is not a true soil-dwelling species but lives in the organic material (litter-dwelling organism) and occurs

primarily on sites rich in organic matter. Nevertheless, *Eisenia foetida* is considered to be a representative test species since its sensitivity is comparable to that of soil-dwelling earthworms. *Lumbricus terrestris* or *Aporrectodea caliginosa* would provide a higher ecological relevance; however, these soil indigenous species are more difficult to handle in the laboratory and breeding is more time consuming as their reproduction rate is low (Kula & Larink 1998).

Earthworm – reproduction test

Test organism: *Eisenia foetida* or *Eisenia andrei* (red worm).
Principle/Procedure: The production of cocoons and juveniles of adult worms exposed to potentially contaminated soil is determined.
Exposure time: After 4 weeks, adult worms are removed from soil, and cocoons are counted. After 8 weeks, juveniles are counted.
Specific equipment: See acute test.
Test parameter(s): number of cocoons and offspring.
Limitations: See acute test.
References/Guidelines: ISO 11268–2 (1998).
Comments: See acute test. Locating and counting of juveniles may be laborious. Application of ascending heat (water bath) should be a useful procedure to dislodge juveniles from soil.

Collembola – reproduction test

Test organism: *Folsomia candida* (springtail).
Principle/Procedure: The effect of potentially contaminated soil on the reproduction of springtails is determined.
Exposure time: 4 weeks.
Test parameter(s): mortality and number of offspring.
References/Guidelines: ISO 11267 (1999); Wiles and Krogh (1998).
Comments: Springtails play an important ecological role in the recycling of nutrients in many ecosystems. Moreover, they are prey for a variety of invertebrates such as mites, centipedes, spiders, and carabid and rove beetles (Wiles & Krogh 1998). Since breeding and maintenance of *Folsomia candida* in the laboratory is comparatively simple, the test is now well established in terrestrial ecotoxicity testing.

Enchytreid – reproduction test

Test organism: *Enchytraeus albidus*, *Enchytraeus* sp. (pot worm).
Principle/Procedure: Determination of the effect on reproduction of enchytreids exposed to potentially contaminated soil.
Exposure time: 6 weeks.
Test parameter(s): mortality and number of offspring.

References/Guidelines: Römbke *et al.* (1999); Rundgren and Augustsson (1998). ISO/DIS 16387 (2002): Effects of soil pollutants on Enchytraeidae: Determination of effects on reproduction.

Comments: Besides Lumbricidae, Enchytraeidae belong to the most important oligochaete families in Europe (Rundgren & Augustsson 1998). Enchytraeids, also called pot worms, are little white worms that feed on bacteria and fungi and ingest dead organic matter. In the laboratory, enchytraeids are easy to culture and maintain.

Ostracod test

A rather new testkit (OSTRACODTOXKIT F™, MicroBioTests Inc., Nazareth, Belgium) is available for short-term direct contact invertebrate testing. Neonates of the benthic ostracod crustacean *Heterocypris incongruens* are used as test organisms. Mortality and growth inhibition resulting from direct contact with contaminated material is determined after 6 days.

The test was originally developed for direct contact sediment testing but efforts are made to evaluate the method for its application to soil. The advantage of the test is its availability as testkit including the availability of standardised biological material. Furthermore, the test duration is rather short compared to other invertebrate tests.

Plant Tests
Effects on seedling germination, emergence and early growth of terrestrial plants.

Test species: A multitude of test species are suggested in guidance papers, 27 species mentioned in four selected methods are listed in Table 10.2, an extended list comprising 75 species can be found at GCPF (2000). The number of species recommended usually ranges from 2 to 10 species, and monocotyledon and dicotyledon species are selected in varying ratios. OECD 208 (1984) recommends the selection of at least one of each category (a category comprises the species with the same number in parenthesis following OECD) from the following table. In ISO 11269-2 (1995), the application of at least one from the monocotyledonous and one from the dicotyledonous species is proposed, whereas US FIFRA and ASTM suggest a ratio of mono- to dicotyledon species of 2:3. With respect to 'evaluation of the soil as a habitat for plant growth' only four species comprising oat, wild turnip, cress and bean were proposed (DECHEMA 2001).

Principle/Procedure: The suppression of germination, emergence and early plant growth resulting from contact to potentially contaminated soil or aqueous soil extracts is assessed.

Exposure time: depending on species (up to 28 days).

Specific equipment: glasshouse, plant growth chamber.

Test parameter(s): inhibition of germination, emergence and growth (biomass).

Table 10.2 Plant species for toxicity testing recommended by international organisations

Plant	Species	Family	Guidelines*
Dicotyledonae			
Lettuce	*Lactuca sativa*	Compositae (Asteraceae)	OECD (3), ISO, ASTM
Chinese cabbage	*Brassica campestris* var. *chinensis*	Cruciferae (Brassicaceae)	OECD (2), ISO, ASTM
Cress	*Lepidium sativum*	Cruciferae (Brassicaceae)	OECD (3), ISO, ASTM
Mustard	*Brassica alba*	Cruciferae (Brassicaceae)	OECD (2), ISO, ASTM
Radish	*Raphanus sativus*	Cruciferae (Brassicaceae)	OECD (2), ISO, ASTM
Rape	*Brassica napus*	Cruciferae (Brassicaceae)	OECD (2), ISO, ASTM, US FIFRA
Turnip	*Brassica rapa*	Cruciferae (Brassicaceae)	OECD (2), ISO, ASTM
Wild cabbage	*Brassica oleracea*	Cruciferae (Brassicaceae)	ASTM
Cucumber	*Cucumis sativa*	Cucurbitaceae	ASTM, US FIFRA
Fenugreek	*Trifolium ornithopodioides*	Leguminosae (Fabaceae)	OECD (3), ISO, ASTM
Mung bean	*Phaseolus aureus*	Leguminosae (Fabaceae)	OECD (3), ASTM
Bean	*Phaseolus vulgaris*	Leguminosae (Fabaceae)	ASTM
Red clover	*Trifolium pratense*	Leguminosae (Fabaceae)	OECD (3), ASTM
Soybean	*Glycine max* (*G. soja*)	Leguminosae (Fabaceae)	ASTM, US FIFRA
Vetch	*Vicia sativa*	Leguminosae (Fabaceae)	OECD (3), ASTM
Tomato	*Lycopersicon esculentum*	Solanaceae	ISO, US FIFRA
Carrot	*Daucus carota*	Umbelliferae (Apiaceae)	ASTM
Monocotyledonae			
Barley	*Hordeum vulgare*	Gramineae (Poaceae)	ISO
Oat	*Avena sativa*	Gramineae (Poaceae)	OECD (1), ISO, ASTM
Rice	*Oryza sativa*	Gramineae (Poaceae)	OECD (1), ISO, ASTM
Rye	*Secale cereale*	Gramineae (Poaceae)	ISO
Ryegrass	*Lolium perenne*	Gramineae (Poaceae)	OECD (1), ISO, ASTM, US FIFRA
Sorghum	*Sorghum bicolor*	Gramineae (Poaceae)	OECD (1), ISO, ASTM
Sweetcorn	*Zea mays*	Gramineae (Poaceae)	ISO, ASTM, US FIFRA
Wheat	*Triticum aestivum*	Gramineae (Poaceae)	OECD (1), ISO, ASTM, US FIFRA
Onion	*Allium cepa*	Liliaceae	US FIFRA

*OECD 208 (1984), ISO 11269-2 (1995), ASTM E1598 (1994), US FIFRA (1982); OECD recommends the selection of at least one species of each of the three categories as indicated by numbers in parenthesis whereas ISO, ASTM and US FIFRA suggest the selection of monocotyledonous and dicotyledonous species in varying ratios.

Limitations: It has to be considered that dissipation of volatile substances and degradation due to biological activity during the exposure time influence the result of the toxicity test. The selection of an appropriate reference (control) soil is crucial for obtaining correct results, since the effect measured in contaminated soils is in general related to the reference.

References/Guidelines: ISO 11269-2 (1995); OECD 208 (1984, 2000); ASTM E1598 (1994); US FIFRA (1982); Wang and Freemark (1995).

Comments: While initially the prevention of adverse effects of herbicides on economically valuable plants was a major task, current plant toxicity testing concentrates on the assessment of chemicals in general. However, the selection of test species included in guidelines is still predominantly based on crop species representing a limited number of families. Assessment of soil that serves as habitat for plants may require ecologically important species other than annual crops.

Inhibition of root elongation

Test organism: *Hordeum vulgare* (barley), *Latuca sativa* (lettuce), *Pancum miliaceum* (millet seeds), *Lycopersicon esculentum* (tomato), *Cucumis sativus* (cucumber), *Glycine max* (soybean), *Brassica oleracea* (cabbage), *Avena sativa* (oat), *Lolium perenne* (perennial ryegrass), *Allium cepa* (common onion), *Daucus carota* (carrot), *Zea mays* (corn).
Principle/Procedure: Root lengths of the seeds grown in potentially contaminated soil or aqueous soil extracts are measured.
Exposure time: depending on species (e.g. 7 days for barley).
Specific equipment: glasshouse, plant growth chamber.
Test parameter(s): root length.
References/Guidelines: ISO 11269-1 (1993); USEPA OPPTS 850.4200 (1996); Wang (1987).
Comments: The ISO guideline focuses only on barley (*Hordeum vulgare* L.) as a test species.

Microbial Tests
Single species tests (addition of test strains)

Microtox™ Solid-phase test SDI: *Vibrio fischeri* is added to a suspension of the test sample. Subsequently, the mixture is filtered using the device supplied with the kit, and the light emission of the supernatant is determined. The method is a further development of the acute luminescent bacteria test for aqueous samples.
Contact test with *Bacillus cereus* DSM No. 351: Effects on the dehydrogenase activity of the test bacterium after an exposure time of 2 hours are investigated using resazurine reduction (Rönnpagel *et al.* 1995).
Toxi-ChromoPad™ (EBPI, Brampton, Ontario, Canada): This pad toxicity bioassay determines acute toxicity in solid samples with a mutant of the bacterium *Escherichia coli*. The assay allows the bacteria to grow in direct contact with the toxicants in the sample.
  This assay is closely related to bioassays reported by Paton *et al.* (1996a,b) and Sousa *et al.* (1998) where water extracts of soil samples are exposed over short periods to lux-marked bacteria, including *Escherichia coli*, *Rhizobium* and a range of pseudomonads. The marking involves placing the lux genes downstream of a strong constitutive promoter and so ensures that light output from the cells report on overall metabolic activity, and so contaminant

toxicity is related to the decline in luminescence of environmentally relevant bacteria.

Microbial communities test (indigenous populations)

Soil respiration
Test organism: aerobic and facultative anaerobic microorganisms in the soil.
Principle/Procedure: The substrate-induced microbial respiratory activity of the soil is determined.
Exposure time: up to 24 hours.
Specific equipment: A device which provides continuous supply of O_2 is preferable.
Test parameter(s): O_2 uptake or CO_2 production.
References/Guidelines: ISO 14240-1 (1997).

Ammonium oxidation

Test organism: autotrophic ammonium oxidising bacteria in soil.
Principle/Procedure: Ammonium sulfate is added to the soil serving as a substrate for ammonium oxidising bacteria. The accumulation rate of nitrite is determined and used as an estimate for the activity.
Exposure time: 6 hours.
Test parameter(s): potential ammonium oxidation rate.
Limitations: Indigenous ammonium oxidising bacteria may have already adapted to the conditions in contaminated soil. A possible underestimation of toxicity should be considered when interpreting test results.
References/Guidelines: ISO/DIS 15685 (2001); DECHEMA (1995).

Bioassays using aqueous extracts (elutriates and leachates)

Algae tests
Freshwater algal growth inhibition test

Test organism: *Selenastrum capricornutum* ATCC 22662 (renamed *Raphidocelis subcapitata* and more recently *Pseudokirchneriella subcapitata*), *Scenedesmus subspicatus* 86.81 SAG, *Chloressa vulgaris* CCAP 211/11b.
Principle/Procedure: The growth inhibition of alga exposed to soil elutriates is determined.
Exposure time: 72, 96 hours.
Testkits: ALGALTOXKIT F™ (MicroBioTests Inc., Nazareth, Belgium).
Specific equipment: device for determination of cell concentration (e.g. electronic particle counter, microscope with counting chamber, fluorimeter, spectrophotometer, colorimeter).
Test parameter(s): cell concentration, biomass (cell counting, optical density, fluorescence).

Limitations: Samples with high concentrations of nutrients may lead to a promotion of growth, and toxic effects of pollutants may be masked. Coloured samples may influence the growth due to reduced light intensity and limit the use of optical density and fluorescence as a parameter for cell concentration.

References/Guidelines: OECD 201 (1984); Environment Canada EPS 1/RM/25 (1992b), USEPA OPPTS 850.5400 (1996).

Comments: Environment Canada (1992b) has outlined a test protocol for a miniaturised alga growth inhibition assay. The application of 96-well microtiter plates facilitates the measurement of a high number of samples.

Plant tests
Lemna minor growth inhibition test

Test organism: *Lemna minor* (duckweed).
Principle/Procedure: The effect of soil elutriates on growth is determined.
Exposure time: 4 days, 7 days.
Test parameter(s): growth rate.
References/Guidelines: ASTM E1415-91 (1998); USEPA OPPTS 850.4400 (1996).

Invertebrate tests
Immobilisation test with *Daphnia magna*

Test organism: *Daphnia magna* (water flea).
Principle/Procedure: The mortality of the crustacean exposed to soil elutriate or leachate is determined.
Exposure time: 24, 48 hours.
Testkits: DAPHTOXKIT F™ MAGNA, MicroBioTests Inc., Nazareth, Belgium
Specific equipment: temperature controlled at 20°C.
References/Guidelines: OECD 202 (1984).

Invertebrate testkits

Several short-term invertebrate tests for aqueous media, commercially available as testkits (MicroBioTests Inc., Nazareth, Belgium), have been developed by the research teams of Prof. Dr G. Persoone at the Laboratory for Biological Research in Aquatic Pollution (LABRAP) at the University of Ghent in Belgium. These TOXKITS contain all necessary materials, in particular the test organisms in *dormant* or *immobilised* form.

CERIODAPHTOXKIT F™ – *Ceriodaphnia dubia* (mortality after 24 hours).
THAMNOTOXKIT F™ – *Thamnocephalus platyurus* (mortality after 24 hours).
ROTOXKIT F™ and ROTOXKIT F™ short chronic – *Brachionus calyciflorus* (mortality and reproduction after 24 and 48 hours).
PROTOXKIT F™ – *Tetrahymena thermophila* (biomass after 24 hours).

Microbial tests
Vibrio fischeri – luminescent bacteria test – acute

Test organism: *Vibrio fischeri* (NRRL-B 11177).
Principle/Procedure: The inhibitory effects of potentially contaminated soil elutriates on bacterial luminescence are measured.
Exposure time: 5–30 minutes.
Testkits: Microtox® (Azur Environmental/SDI, formerly Microbics Corporation, USA), LUMIStox™ (Dr Lange, Germany), Biotox™ (Aboatox, Finland), ToxAlert™ (Merck, Germany).
Specific equipment: luminometer, cooling device (15°C).
Test parameter(s): percent bioluminescence inhibition compared to a 2% NaCl solution or a reference elutriate.
Limitations: highly coloured and turbid elutriates.
References/Guidelines: ISO 11348 (1998); Environment Canada EPS 1/RM/24 (1992a).
Comments: Due to the sensitivity of *Vibrio fischeri* to a broad range of pollutants, the reliability, reproducibility and the rather short duration of the test, this method is frequently used for diverse investigations. Since *Vibrio fischeri* is a marine bacterium, it lacks ecological relevance for soil, however, it is often used as a screening test.

The determination of the endpoint is – like for all other tests based on a photometric measurement – influenced by colour and turbidity of the sample. In this case, the use of a double chamber cuvette enables the estimation of the impact of colour or turbidity. Another method utilises the optical density of the sample for the calculation of a colour-corrected inhibition value.

A promising flash method was proposed by Lappalainen *et al.* (1999), applying a new measurement technique. Bioluminescence is determined immediately after *Vibrio fischeri* is exposed to the soil slurry and again after an incubation period of less than one minute. The first value serves as reference, assuming that the bacteria are not yet affected by the toxicants in the sample. The second value, expressed relative to the first value, indicates the effect of the toxicant. This procedure provides an alternative for coloured and turbid samples.

Vibrio fischeri – luminescent bacteria test – chronic

Test organism: *Vibrio fischeri* (NRRL-B 11177).
Principle/Procedure: Inhibitory effects of potentially contaminated soil elutriates on bacterial luminescence (and growth) are measured after chronic exposure.
Exposure time: 22–24 hours.
Testkits: Microtox® (SDI), LUMIStox (Dr Lange, Germany).

Specific equipment: luminometer, cooling device (15 and 27°C for LUMIStox and Microtox®, respectively).

Test parameter(s): percent bioluminescence inhibition (Microtox®, LUMIStox) and bacterial growth (LUMIStox) compared to a 2% NaCl solution or a reference elutriate.

Limitations: see acute test.

Genotoxicity tests

The application of a metabolic activation system is recommended, since pollutants such as PAH are not genotoxic in their parent chemical form but are transformed during metabolism by enzymes of organisms. A commonly used system is the S-9 homogenate prepared from the liver of rats treated with enzyme inducing agents.

Ames-test

Test organism: *Salmonella typhimurium* TA100 and TA98.

Principle/Procedure: The mutagenic potential of aqueous soil extracts is determined. The Ames-test is based on the fact that autotrophic mutants of the bacterium carry a defect gene, making it unable to synthesise histidine. In the presence of mutagenic agents, the mutation can be reversed with the gene regaining its function. After incubation of the strain with the suspected mutagen in the medium lacking histidine, the number of revertant colonies that are able to grow on the agar is counted. Optionally, rat liver enzymes (S-9 homogenate) can be applied to convert materials without inherent genotoxic activity to active genotoxins.

Exposure time: 48 hours.

Testkits: Muta-ChromoPlate™ (a microplate version of the traditional 'pour plate' Ames test, Environmental Bio-detection Products Incorporated (EBPI), Brampton, Canada).

Test parameter(s): effect of the sample on the number of revertants.

References/Guidelines: DIN 38415-4 (1999).

Umu-test

Test organism: *Salmonella choleraesius* subsp. *chol.* TA1535/pSK1002 (formerly *Salmonella typhimurium*).

Principle/Procedure: The mutagenic potential of aqueous soil extracts is determined. Spontaneous activity of the umu C gene is indicated by the activity of galactosidase as an indicator.

Exposure time: 2 hours.

Test parameter(s): galactosidase activity.

References/Guidelines: DIN 38415-3 (1996); ISO 13829 (2000).

Vitotox Test

Test organism: two recombinant *Salmonella typhimurium* TA104 test strains.
Principle/Procedure: The genotoxicity and cytotoxicity of aqueous soil extracts are determined. In the presence of DNA damaging compounds, the initiation of a cascade of reactions known as the SOS response will take place leading to the activation of the luciferase gene. Light emission is measured as a function of the activity of the luciferase gene indicating genotoxicity.
Exposure time: up to 4 hours.
Testkits: Vitotox® (ThermoLabsystems, Helsinki, Finland).
Test parameter(s): light emission as an indicator for the activity of the luciferase reporter gene system as a result of the SOS response initiated by DNA damage.

SOS-Chromotest

Test organism: *Escherichia coli* PQ37.
Principle/Procedure: The genotoxic potential of soil elutriates and extracts is determined. The microplate genotoxicity bioassay is based on the primary response of genetically engineered bacteria to determine DNA damage. The SOS promoter does not activate the SOS system but induces the synthesis of galactosidase, which catalyses the formation of colour. Furthermore, alkaline phosphatase activity is measured to indicate non-genotoxic inhibition of *Escherichia coli*. Missing phosphatase activity indicates dead or inactive cells that are not able to respond to genotoxic substances. The test can be performed with and without S-9 activation.
Exposure time: approximately 3 hours.
Testkits: SOS-Chromotest (Environmental Bio-detection Products Incorporated (EBPI), Brampton, Canada).
Specific equipment: microplate reader, incubation chamber (37°C).
Test parameter(s): formation of colour due to β-galactosidase activity induced by DNA damage.

Mutatox Test

Test organism: *Vibrio fischeri*, strain M169.
Principle/Procedure: Genotoxic effects of aqueous soil extracts are measured. The strain is a dark variant of the luminescent bacteria and poorly emits light during growth cycles. In the presence of genotoxic materials, light production is restored.
Exposure time: 24 hours.
Testkits: Mutatox® (Azur Environmental, formerly Microbics Corporation, USA).
Test parameter(s): light emission indicating the presence of genotoxic agents.

Abbreviations

ASTM	American Society for Testing and Materials
BTEX	benzene, toluene, ethylbenzene and xylenes
CARACAS	concerted action on risk assessment for contaminated sites
CEN	Comité Européen de Normalisation (European Committee for Standardisation)
CERCLA	Comprehensive Environmental Response, Compensation, and Liability Act (USA)
CLARINET	Contaminated Land Rehabilitation Network for Environmental Technologies
CSTEE	Scientific Committee on Toxicity, Ecotoxicity and the Environment of the EU
DECHEMA	Deutsche Gesellschaft für chemisches Apparatewesen, Chemische Technik und Biotechnologie (Society for Chemical Engineering and Biotechnology)
DIN	Deutsches Institut für Normung (German Institute for Standardisation)
EC_{50}	median effective concentration
EU	European Union
IC_{50}	inhibiting concentration for a 50% effect
ISO	International Organization for Standardization
LC_{50}	median lethal concentration
LOEC	lowest observed effect concentration
$\log K_{ow}$	logarithm of the octanol-water partitioning coefficient
LUFA	Landwirtschaftliche Untersuchungs- und Forschungsanstalt Speyer (Agricultural Testing and Research Station Speyer)
NOEC	no observed effect concentration
OECD	Organisation for Economic Cooperation and Development
OENORM	Österreichisches Norm (Austrian Standard)
PAH	polycyclic aromatic hydrocarbons
PEC	predicted environmental concentration
PNEC	predicted no effect concentration
USEPA	United States Environmental Protection Agency
US FIFRA	United States Federal Insecticide, Fungicide, and Rodenticide Act

References

ASTM E1415-91 (1998) Standard guide for conducting static toxicity test with *Lemna gibba* G3. American Society for Testing and Materials, West Conshohocken.
ASTM E1598-94 (1994) Standard practice for conducting early seedling growth tests. American Society of Testing and Materials, West Conshohocken.

ATSDR (2001) CERCLA Priority List of Hazardous Substances. Agency for Toxic Substances and Disease Registry, U.S. Department of Health and Human Services in cooperation with the U.S. Environmental Protection Agency, Washington, DC.

Becker, D.S. & Ginn, T. (1990) Effects of sediment holding time on sediment toxicity. Prepared for the U.S. EPA, Region 10, Office of Puget Sound, PTI Environmental Services, Bellevue, WA.

CSTEE (2000) Opinion on 'The Available Scientific Approaches to Assess the Potential Effects and Risk of Chemicals on Terrestrial Ecosystems', (online). 2000-11-09, http://europa.eu.int/comm/food/fs/sc/sct/out83_en.pdf, accessed at 2002-02-28.

Cuypers, C., Grotenhuis, T., Joziasse, J. & Rulkens, W. (2000) Rapid persulfate oxidation predicts PAH bioavailability in soils and sediments. *Environmental Science and Technology*, **34**(10), 2057–2063.

DECHEMA (1995) *Bioassay for soils*, Ad-Hoc-Committee "Methods for toxicological/ecotoxicological assessment of soils", Deutsche Gesellschaft für chemisches Apparatewesen, Chemische Technik und Biotechnologie e.V., G. Kreysa and J. Wiesner, Frankfurt am Main.

DECHEMA (2001) Biologische Testverfahren für Boden und Bodenmaterial, Arbeitsgruppe "Validierung biologischer Testmethoden für Böden", Deutsche Gesellschaft für chemisches Apparatewesen, Chemische Technik und Biotechnologie e.V., Dott, W., Frankfurt am Main.

dela Cruz, M.A.T., Loibner, A.P., Szolar, O.H.J., Leyval, C., Joner, E. & Braun, R. (2000) Ecotoxicity monitoring of phytoremediation processes, In *Contaminated Soil* 2000, W. Harder and F. Arendt (eds), Thomas Telford Publishing, London, UK, **1**, 351–354.

DIN 38414-4 (1999) Determination of leachability by water. German standard methods for the examination of water, waste water and sludge; sludge and sediments (group S) – Part 4, Berlin.

DIN 38415-3 (1996) Determination of the genotype potential of water and waste water components with the umu-test. German standard methods for the examination of water, waste water and sludge – Sub-animal testing (group T) – Part 3, Berlin.

DIN 38415-4 (1999) Determination of the genotoxic potential using the *Salmonella* microsome test (Ames Test). German standard methods for the examination of water, waste water and sludge – Sub-animal testing (group T) – Part 4, Berlin.

Environment Canada EPS 1/RM/24 (1992a) Biological test method: toxicity test using luminescent bacteria (*Photobacterium phosphoreum*). Environmental Protection, Conservation and Protection, Environmental Protection Series Report, Ottawa.

Environment Canada EPS 1/RM/25 (1992b) Biological test method: growth inhibition test using the freshwater alga *Selenastrum capricornutum*. Environmental Protection, Conservation and Protection, Environmental Protection Series Report, Ottawa.

Environment Canada EPS 1/RM/34 (1999) Guidance document on application and interpretation of single-species tests in environmental toxicology, Environmental Protection Series, Ottawa.

European Commission (2001) White Paper on 'Strategies for a Future Chemicals Policy', (online). 2001-02-27, http://europa.eu.int/eur-lex/en/com/wpr/2001/com2001_0088en01.pdf, accessed at 2002-02-28.

Ferguson, C., Darmendrail, D., Freier, K., Jensen, B.K., Jensen, J., Kasamas, H., Urzelai, A. & Vegter, J. (eds) (1998) *Risk Assessment for Contaminated Sites in Europe*, **1**, Scientific Basis, LQM press, Nottingham.

Ferguson, C. & Kasamas, H. (eds) (1999) *Risk Assessment for Contaminated Sites in Europe*, **2**, Policy Frameworks, LQM press, Nottingham.

Forbes, V.E. & Forbes, T.L. (1994) *Ecotoxicology in Theory and Practice*, Chapman & Hall, London.

Gallo, M.A. & Doull, J. (1991) Principles of toxicology, in *Casarett and Doull's Toxicology: The Basic Science of Poisons*, 4th edn, M.O. Amdur, J. Doull and C.D. Klaassen (eds), Pergamon Press, New York, pp. 3–11.

Gaudet, C. (1994) A framework for ecological risk assessment at contaminated sites in Canada: review and recommendations, Environment Canada, Scientific Series No. 199, Cat. No. En 36-502/199E, Ottawa.

GCPF (2000) *A Comparative Review of Terrestrial (non-target) Plant Test Methods*. Background document, Global Crop Protection Federation.

Hoffman, D.J., Rattner, B.A., Burton, G.A. Jr & Cairns, J. Jr (eds) (1995) *Handbook of Ecotoxicology*, Lewis Publishers, an imprint of CRC Press, Boca Raton.

ISO 11267 (1999) Inhibition of reproduction of Collembola (*Folsomia candida*) by soil pollutants. ISO, Geneva.

ISO 11268-1 (1993) Effects of pollutants on earthworms (*Eisenia foetida*). Part 1: Determination of acute toxicity using artificial soil substrate. ISO, Geneva.

ISO 11268-2 (1998) Effects of pollutants on earthworms (*Eisenia foetida*). Part 2: Determination of effects on reproduction. ISO, Geneva.

ISO 11269-1 (1993) Determination of the effects of pollutants on soil flora. Part 1: Method for the measurement of inhibition of root growth. ISO, Geneva.

ISO 11269-2 (1995) Determination of the effects of pollutants on soil flora. Part 2: Effects of chemicals on the emergence and growth of higher plants. ISO, Geneva.

ISO 11348 (1998) Water quality – Determination of the inhibitory effect of water samples on the light emission of *Vibrio fischeri* (Luminescent bacteria test). Part 1–3. ISO, Geneva.

ISO 13829 (2000) Water quality – Determination of the genotoxicity of water and waste water using the umu-test. ISO, Geneva.

ISO 14240-1 (1997) Determination of soil microbial biomass. Part 1: Substrate-induced respiration method. ISO, Geneva.

ISO/DIS 15685 (2001) Soil quality – Determination of potential nitrification – Rapid test by ammonium oxidation. ISO, Geneva.

ISO/FDIS 15799 (2001) Soil quality – Guidance on the ecotoxicological characterization of soils and soil materials. ISO, Geneva.

Keddy, C., Greene, J.C. & Bonell, M.A. (1994) A review of whole organism bioassays for assessing the quality of soil, freshwater sediment and freshwater in Canada. Environment Canada, Scientific series No. 198, Cat. No. En 36-502/198-E, Ottawa.

Klaassen, C.D. & Eaton, D.L. (1991) Principles of toxicology, in *Casarett and Doull's Toxicology: The Basic Science of Poisons*, 4th edn, M.O. Amdur, J. Doull and C.D. Klaassen (eds), Pergamon Press, New York, pp. 12–49.

Kula, H. & Larink, O. (1998) Tests on the Earthworms *Eisenia fetida* and *Aporrectodea caliginosa*, in *Handbook of soil invertebrate toxicity tests*, H. Lokke and C.A.M. van Gestel (eds), John Wiley & Sons, Chichester, pp. 95–112.

Lappalainen, J., Juvonen, R., Vaajasaari, K. & Karp, M. (1999) A new flash method for measuring the toxicity of solid and coloured samples. *Chemosphere*, **38**(5), 1069–1083.

Liste, H.-H. & Alexander, M. (2002) Butanol extraction to predict bioavailability of PAHs in soil. *Chemosphere*, **46**, 1011–1017.

Loibner, A.P., Holzer, M., Gartner, O., Szolar, O.H.J. & Braun, R. (2000) The use of sequential supercritical fluid extraction for bioavailability investigations of PAH in soil. *Die Bodenkultur*, **51**(4), 225–233.

Luthy, R.G., Aiken, G.R., Brusseau, M.L., Cunningham, S.D., Gschwend, P.M., Pignatello, J.J., Reinhard, M., Traina, S.J., Weber Jr, W.J. & Westall, J.C. (1997) Sequestration of hydrophobic organic contaminants by geosorbents. *Environmental Science and Technology*, **31**(12), 3341–3347.

Moore, D.R.J. & Caux, P.-Y. (1997) Estimation low toxic effects. *Environmental Toxicology and Chemistry*, **16**(4), 794–801.

OECD 201 (1984) Guideline for testing of chemicals: alga, growth inhibition test. Organization for Economic Cooperation and Development, Paris.

OECD 202 (1984) Guideline for testing of chemicals: *Daphnia* sp., acute immobilisation test and reproduction test, Part I. Organization for Economic Cooperation and Development, Paris.

OECD 207 (1984) Guideline for testing of chemicals: earthworm, acute toxicity tests. Organization for Economic Cooperation and Development, Paris.

OECD 208 (1984) Guideline for testing of chemicals: terrestrial plants, growth test. Organization for Economic Cooperation and Development, Paris.

OECD 208 (2000) Guideline for the testing of chemicals. Proposal for updating guideline 208: Terrestrial (non-target) plant test: 208A: Seedling emergence and seedling growth test. 208B: Vegetative vigour test. (Draft document). Organization for Economic Cooperation and Development, Paris.

Paton, G.I., Campbell, C.D., Cresser, M.S., Glover, L.A. & Killham, K. (1996a) Use of genetically modified biosensors for soil ecotoxicity testing. Bioindicators of Soil Health. C.F. Parkhurst, B.M. Doube, V.V.S.R. Gupta (eds), CAB International, Wallingford, pp. 397–418.

Paton, G.I., Palmer, G., Burton, M., Rattray, E.A.S., Glover, L.A., McGrath, S. & Killham, K. (1996b) Development of an acute and chronic ecotoxicity assay using lux-marked *Rhizobium leguminosarum* bv. *trifolii*. FEMS letters in Applied Microbiology, **24**, 296–300.

Persoone, P., Janssen, C. & De Coen, W. (eds) (2000) *New microbiotests for routine toxicity screening and biomonitoring.* Kluwer Academic/Plenum Publishers, New York.

Römbke, J. & Moltmann, J.F. (1996) *Applied Ecotoxicology.* Lewis Publishers, an imprint of CRC press, Boca Raton.

Römbke, J. *et al.* (1999) Organization and performance on an international ringtest for the validation of the Enchytraeid reproduction test. *UBA Texte* 4/99, Berlin.

Rönnpagel, K., Liß, W. & Ahlf, W. (1995) Microbial bioassays to assess the toxicity of solid-associated contaminants. *Ecotoxicology and Environmental Safety*, **31**, 99–103.

Rost, H., Loibner, A.P., Hasinger, M., Braun, R. & Szolar, O.H.J. (2002) Behaviour of PAHs during cold storage of historically contaminated soil samples. *Chemosphere*, **49**(10), 1239–1246.

Rundgren, S. & Augustsson, A.K. (1998) Tests on the Enchytraeid *Cognettia sphagnetorum* (Vejdovsky) 1877, in *Handbook of Soil Invertebrate Toxicity Tests*. H. Lokke and C.A.M. Van Gestel (eds), John Wiley & Sons, Chichester, pp. 73–94.

Rutgers, M. & Breure, A.M. (1999) Risk assessment, microbial communities, and pollution-induced community tolerance, *Human and Ecological Risk Assessment*, **5**(4), 661–670.

Sousa, S., Duffy, C., Weitz, H., Glover, L.A., Henkler, R. & Killham, K. (1998) Use of a lux bacterial biosensor to identify constraints to remediation of contaminated environmental samples. *Environmental Toxicology and Chemistry*, **17**, 1039–1045.

Suter, G.W., II (1993) *Ecological Risk Assessment*, Lewis Publishers, Boca Raton.

Suter, G.W., II, Efroymson, R.A., Sample, B.E. & Jones, D.S. (2000) *Ecological Risk Assessment for Contaminated Sites*, Lewis Publishers, an imprint of CRC Press, Boca Raton.

US FIFRA (1982) FIFRA, 40CFR, Part 158.540, Subdivision J, Parts 122–1 and 123–1.

USEPA (1992) Framework for ecological risk assessment. Risk Assessment Forum, EPA/630/R-92/001, Washington, DC.

USEPA (1998) Guidelines for ecological risk assessment. Risk Assessment Forum, EPA/630/R-95/002F, Washington, DC.

USEPA OPPTS 850.4200 (1996) Ecological Effect Test Guidelines, 850 Series (Proposal): Seed Germination/Root Elongation Toxicity Test.

USEPA OPPTS 850.4400 (1996) Ecological Effect Test Guidelines, 850 Series (Public Draft): Aquatic Plant Toxicity Test Using *Lemna* spp.

USEPA OPPTS 850.5400 (1996) Ecological Effect Test Guidelines, 850 Series (Public Draft): Algal Toxicity, Tiers I and II.

Wang, W. (1987) Root elongation method for toxicity testing of organic and inorganic pollutants, *Environmental Toxicology and Chemistry*, **6**, 409–414.

Wang, W. & Freemark, K. (1995) The use of plants for environmental monitoring and assessment. *Ecotoxicology and Environmental Safety*, **30**, 289–301.

Wenzel, W.W. & Blum, W.E.H. (1999) Effect of sampling, sample preparation and extraction techniques on mobile metal fractions in soils, in D.C. Adriano, Z.-S. Chen, S.-S. Yang, and I.K. Iskandar (eds), *Biogeochemistry of trace metals*, Advances in Environmental Sciences, Science Reviews, Northwood, pp. 121–172.

Wiles, J.A. & Krogh, P.H. (1998) Tests with the Collembolans *Isotoma viridis, Folsomia candida* and *Folsomia fimetaria*, in *Handbook of Soil Invertebrate Toxicity Tests*. H. Lokke and C.A.M. Van Gestel (eds), John Wiley & Sons, Chichester, pp. 131–156.

Appendix 1

ISO – International Organization for Standardization Committee: TC 190 Soil quality: Published standards (November 2002)

ISO 10381-2:2002 Soil quality – Sampling – Part 2: Guidance on sampling techniques.

ISO 10381-3:2001 Soil quality – Sampling – Part 3: Guidance on safety.

ISO 10381-6:1993 Soil quality – Sampling – Part 6: Guidance on the collection, handling and storage of soil for the assessment of aerobic microbial processes in the laboratory.

ISO 10382:2002 Soil quality – Determination of organochlorine pesticides and polychlorinated biphenyls – Gaschromatographic method with electron capture detection.

ISO 10390:1994 Soil quality – Determination of pH.

ISO 10573:1995 Soil quality – Determination of water content in the unsaturated zone – Neutron depth probe method (available in English only).

ISO 10693:1995 Soil quality – Determination of carbonate content – Volumetric method.

ISO 10694:1995 Soil quality – Determination of organic and total carbon after dry combustion (elementary analysis).

ISO/TR 11046:1994 Soil quality – Determination of mineral oil content – Method by infrared spectrometry and gas chromatographic method.

ISO 11047:1998 Soil quality – Determination of cadmium, chromium, cobalt, copper, lead, manganese, nickel and zinc – Flame and electrothermal atomic absorption spectrometric methods.

ISO 11048:1995 Soil quality – Determination of water-soluble and acid-soluble sulfate.

ISO 11074-1:1996 Soil quality – Vocabulary – Part 1: Terms and definitions relating to the protection and pollution of the soil.

ISO 11074-2:1998 Soil quality – Vocabulary – Part 2: Terms and definitions relating to sampling.

ISO 11074-4:1999 Soil quality – Vocabulary – Part 4: Terms and definitions related to rehabilitation of soils and sites.

ISO 11259:1998 Soil quality – Simplified soil description.

ISO 11260:1994 Soil quality – Determination of effective cation exchange capacity and base saturation level using barium chloride solution.

ISO 11260:1994/Cor 1:1996.

ISO 11261:1995 Soil quality – Determination of total nitrogen – Modified Kjeldahl method.

ISO 11263:1994 Soil quality – Determination of phosphorus – Spectrometric determination of phosphorus soluble in sodium hydrogen carbonate solution.

ISO 11265:1994 Soil quality – Determination of the specific electrical conductivity.

ISO 11265:1994/Cor 1:1996.

ISO 11266:1994 Soil quality – Guidance on laboratory testing for biodegradation of organic chemicals in soil under aerobic conditions.

ISO 11267:1999 Soil quality – Inhibition of reproduction of Collembola (*Folsomia candida*) by soil pollutants.

ISO 11268-1:1993 Soil quality – Effects of pollutants on earthworms (*Eisenia fetida*) – Part 1: Determination of acute toxicity using artificial soil substrate.

ISO 11268-2:1998 Soil quality – Effects of pollutants on earthworms (*Eisenia fetida*) – Part 2: Determination of effects on reproduction.

ISO 11268-3:1999 Soil quality – Effects of pollutants on earthworms – Part 3: Guidance on the determination of effects in field situations.

ISO 11269-1:1993 Soil quality – Determination of the effects of pollutants on soil flora – Part 1: Method for the measurement of inhibition of root growth.

ISO 11269-2:1995 Soil quality – Determination of the effects of pollutants on soil flora – Part 2: Effects of chemicals on the emergence and growth of higher plants.

ISO 11271:2002 Soil quality – Determination of redox potential – Field method.

ISO 11272:1998 Soil quality – Determination of dry bulk density.

ISO 11274:1998 Soil quality – Determination of the water-retention characteristic – Laboratory methods.

ISO 11276:1995 Soil quality – Determination of pore water pressure – Tensiometer method.

ISO 11277:1998 Soil quality – Determination of particle size distribution in mineral soil material – Method by sieving and sedimentation.

ISO 11277:1998/Cor 1:2002.

ISO 11461:2001 Soil quality – Determination of soil water content as a volume fraction using coring sleeves – Gravimetric method.

ISO 11464:1994 Soil quality – Pretreatment of samples for physico-chemical analyses.

ISO 11465:1993 Soil quality – Determination of dry matter and water content on a mass basis – Gravimetric method.

ISO 11465:1993/Cor 1:1994 (available in English only).

ISO 11466:1995 Soil quality – Extraction of trace elements soluble in aqua regia.

ISO 11508:1998 Soil quality – Determination of particle density.

ISO 13536:1995 Soil quality – Determination of the potential cation exchange capacity and exchangeable cations using barium chloride solution buffered at pH = 8,1.

ISO 13877:1998 Soil quality – Determination of polynuclear aromatic hydrocarbons – Method using high-performance liquid chromatography.

ISO 13878:1998 Soil quality – Determination of total nitrogen content by dry combustion ('elemental analysis').

ISO 14235:1998 Soil quality – Determination of organic carbon by sulfochromic oxidation.

ISO 14238:1997 Soil quality – Biological methods – Determination of nitrogen mineralization and nitrification in soils and the influence of chemicals on these processes.

ISO 14239:1997 Soil quality – Laboratory incubation systems for measuring the mineralization of organic chemicals in soil under aerobic conditions (available in English only).

ISO 14240-1:1997 Soil quality – Determination of soil microbial biomass – Part 1: Substrate-induced respiration method.

ISO 14240-2:1997 Soil quality – Determination of soil microbial biomass – Part 2: Fumigation-extraction method.

ISO 14254:2001 Soil quality – Determination of exchangeable acidity in barium chloride extracts.

ISO 14255:1998 Soil quality – Determination of nitrate nitrogen, ammonium nitrogen and total soluble nitrogen in air-dry soils using calcium chloride solution as extractant.

ISO 14869-1:2001 Soil quality – Dissolution for the determination of total element content – Part 1: Dissolution with hydrofluoric and perchloric acids.

ISO 14869-2:2002 Soil quality – Dissolution for the determination of total element content – Part 2: Dissolution by alkaline fusion.

ISO 14870:2001 Soil quality – Extraction of trace elements by buffered DTPA solution.

ISO 15009:2002 Soil quality – Gas chromatographic determination of the content of volatile aromatic hydrocarbons, naphthalene and volatile halogenated hydrocarbons – Purge-and-trap method with thermal desorption.

ISO 15176:2002 Soil quality – Characterization of excavated soil and other soil materials intended for re-use.

ISO 15178:2000 Soil quality – Determination of total sulfur by dry combustion.

ISO 15473:2002 Soil quality – Guidance on laboratory testing for biodegradation of organic chemicals in soil under anaerobic conditions.

ISO 15709:2002 Soil quality – Soil water and the unsaturated zone – Definitions, symbols and theory.

ISO 15903:2002 Soil quality – Format for recording soil and site information.

Technical Programme TC 190/SC 3 Projects (Chemical methods and soil characteristics): Standards in preparation (November 2002)

ISO/FDIS 11262 Soil quality – Determination of cyanides.

ISO/DIS 11264 Soil quality – Determination of herbicides – Method using HPLC with UV detection.

ISO/DIS 14154 Soil quality – Determination of selected phenols and chloro-
phenols – Gas chromatographic method.
ISO/PRF TS 14256-1 Soil quality – Determination of nitrate, nitrite and
ammonium in field-moist soils by extraction with potassium chloride
solution – Part 1: Manual method.
ISO/DIS 14256-2 Soil quality – Determination of nitrate, nitrite and ammonium
in field-moist soils by extraction with potassium chloride solution – Part 2:
Automated method.
ISO/DIS 14390 Soil quality – Determination of pH.
ISO/FDIS 14507 Soil quality – Pretreatment of samples for determination of
organic contaminants.
ISO/DIS 16703 Soil quality – Determination of mineral oil content by gas
chromatography.
ISO/CD 16720 Soil quality – Determination of dry residue – Method by freezing.
ISO/DIS 16772 Soil quality – Determination of mercury in aqua regia soil
extracts with cold-vapour atomic absorption spectrometry or cold-vapour
atomic fluorescence spectrometry.
ISO/DIS 17380 Soil quality – Determination of total cyanide and easily libera-
table cyanide content – Continuous flow analysis method.
ISO/CD 18287 Soil quality – Determination of polycyclic aromatic hydrocarbons
(PAH) – Gas chromatographic method with mass spectrometric detection.
ISO/DIS 20279 Soil quality – Extraction of thallium and determination by
electrothermal atomic absorption spectrometry.
ISO/CD 20280 Soil quality – Determination of arsenic, antimony and selenium –
Method by extraction in aqua regia and atomic absorption spectrometry.
ISO/CD 22155 Soil quality – Gas chromatographic determination of volatile
aromatic and halogenated hydrocarbons – Static headspace method.
ISO/CD 22892 Soil quality – Guidelines for the identification of target com-
pounds by gas chromatography/mass spectrometry.
ISO/AWI 23161 Soil quality – Identification and determination of organotin
compounds.
ISO/AWI 23470 Soil quality – Determination of effective cation exchange
capacity (CEC) and exchangeable cations using a cobaltihexammine
trichloride solution.

**Technical Programme TC 190/SC 2 Sampling Projects: (Standards in
preparation) (November 2002)**

ISO 10381-1 Soil quality – Sampling – Part 1: Guidance on the design of sampling programmes.
ISO/FDIS 10381-4 Soil quality – Sampling – Part 4: Guidance on the procedure for investigation of natural, near-natural and cultivated sites.

ISO/DIS 10381-5 Soil quality – Sampling – Part 5: Guidance on investigation of soil contamination of urban and industrial sites.

ISO/CD 10381-7 Soil quality – Sampling – Part 7: Guidance on the investigation and sampling of soil gas.

ISO/CD 10381-8 Soil quality – Sampling – Part 8: Guidance on the sampling of stockpiles.

TC 147 Water quality: Explosive standard in preparation.

ISO/CD 22478 Water quality – Determination of selected explosive and related compounds – Method by HPLC with UV detection.

FDIS = Final draft international standard
DIS = Draft international standard
CD = Committee draft
AWI = Approved work item

Appendix 2

Finding out more

Land contamination practice is a relatively young subject and there is much ongoing development in practice, guidance and legislation. This appendix presents some ways of keeping in touch with these developments by exploiting internet and e-mail resources.

United States

The USEPA clean-up information web site provides information about innovative treatment and site characterisation technologies – www.clu-in.org.

TechDirect, hosted by the USEPA's Technology Innovation Office, is an information service that highlights new publications and events of interest to site remediation and site assessment professionals. At the beginning of every month, the service, via e-mail, will distribute a message describing the availability of publications and events. For publications, the message will explain how to obtain a hard copy or how to download an electronic version. You can subscribe at http://www.clu-in.org/newsletters/.

United Kingdom

The following web sites provide or point to useful information:

- Environment Agency – www.environment-agency.gov.uk. Much guidance written by agency staff is available for download. The CLEA pages contain errata, FAQ and software update information.
- DEFRA – www.defra.gov.uk. The CLR 7, 8, 9, 10, TOX and SGV series of reports are available for free download.
- Scottish Executive – www.scotland.gov.uk.
- SEPA – www.sepa.org.uk.
- Scotland and Northern Ireland Forum for Environmental Research (SNIFFER) – www.sniffer.org.uk/.
- CIRIA – www.ciria.org.uk. CIRIA publishes a wide range of technical guidance that is available for purchase.
- CLAIRE – www.claire.co.uk. CLAIRE organises and assesses the results of projects, and disseminates the results to those who have an interest in the remediation of contaminated land.

- FIRST Faraday – http://www.firstfaraday.com/. FIRST is a DTI, EPSRC and NERC funded partnership to encourage academic-industrial collaboration in research, training and technology transfer.

A free e-mail discussion list targeted at local authorities was set up by Land Quality Management at the University of Nottingham in 2000 to facilitate discussion on the development of inspection strategies. It has since evolved into a general contaminated land list. To join the list, send the following message: join contaminated-land-strategies your-first-name your-last-name to jiscmail@jiscmail.ac.uk.

Europe

A number of European networks have been funded to develop and disseminate good practice in land management. The following and their link pages provide a gateway to European sources of information:

- CARACAS – www.caracas.at – focusing on risk assessment and closed in 1998.
- CLARINET – www.clarinet.at – focusing on rehabilitation and closed in 2001.
- NICOLE – www.nicole.org.uk – focusing on industrially contaminated land and began in 1995.
- CABERNET – www.cabernet.org.uk – focusing on brownfield regeneration and began in 2002.

Appendix 3

Contaminated land management: key references used in the UK

ASTM (1995) *Emergency Standard Guide for Risk Based Corrective Action Applied at Petroleum Release Sites.* ASTM Designation E-1739.

ASTM (1998) *Standard Provisional Guide for Risk Based Corrective Action.* ASTM Designation PS104:98.

BRE (2001) Special Digest 1 *Concrete in Aggressive Ground.* BRE, Garston, Watford WD25 9XX.

BRE (1995) Digest 279. *Sulfate and acid attack on concrete in the ground: recommended procedure for soil analysis.* BRE, Garston, Watford WD25 9XX.

British Standards Institution (BS) (1994–2000) BS7755 Soil Quality. BSI London.

BRE (1991) BR212 *Construction of new buildings on gas-contaminated land.* ISBN 085 1255132 BRE, Garston, Watford WO25 9XX.

British Standards Institution (1999) 5930:1999 *Code of practice for site investigations*, BSI (London).

British Standards Institution (2001) BS 10175:2001 *Investigation of Potentially Contaminated Sites.* Code of practice, BSI (London).

CIRIA (1995a) *Report 149 Protecting Development from Methane.* CIRIA 6 Storey's Gate, Westminster, London SW1P 3AU.

CIRIA (1995b) *Report 150 Methane Investigation Strategies.* CIRIA 6 Storey's Gate, Westminster, London SW1P 3AU.

CIRIA (1995c) *Report 151 Interpreting Measurements of Gas in the Ground.* CIRIA 6 Storey's Gate, Westminster, London SW1P 3AU.

CIRIA (1995d) *Report 152 Risk Assessment for Methane and Other Gases from the Ground.* CIRIA 6 Storey's Gate, Westminster, London SW1P 3AU.

DEFRA and Environment Agency (2002a) CLR 7 *Overview of guidance on the assessment of contaminated land.*

DEFRA and Environment Agency (2002b) CLR 8 *Potential contaminants for the assessment of land.*

DEFRA and Environment Agency (2002c) CLR 9 *Contaminants in soils: collation of toxicological data and intake values for humans. Consolidated main report.*

DEFRA and Environment Agency (2002d) CLR 9 TOX1 *Contaminants in soils: collation of toxicological data and intake values for humans. Arsenic.*

DEFRA and Environment Agency (2002e) CLR 9 TOX2 *Contaminants in soils: collation of toxicological data and intake values for humans. Benzo(a)pyrene.*

DEFRA and Environment Agency (2002f) CLR 9 TOX3 *Contaminants in soils: collation of toxicological data and intake values for humans. Cadmium.*

DEFRA and Environment Agency (2002g) CLR 9 TOX4 *Contaminants in soils: collation of toxicological data and intake values for humans. Chromium.*

DEFRA and Environment Agency (2002h) CLR 9 TOX5 *Contaminants in soils: collation of toxicological data and intake values for humans. Inorganic cyanide.*

DEFRA and Environment Agency (2002i) CLR 9 TOX6 *Contaminants in soils: collation of toxicological data and intake values for humans. Lead.*

DEFRA and Environment Agency (2002j) CLR 9 TOX7 *Contaminants in soils: collation of toxicological data and intake values for humans. Mercury.*

DEFRA and Environment Agency (2002k) CLR 9 TOX8 *Contaminants in soils: collation of toxicological data and intake values for humans. Nickel.*

DEFRA and Environment Agency (2002m) CLR 9 TOX10 *Contaminants in soils: collation of toxicological data and intake values for humans. Selenium.*

DEFRA and Environment Agency (2002n) CLR 10 *The Contaminated Land Exposure Assessment Model (CLEA): technical basis and algorithms.*

DEFRA and Environment Agency (2002o) CLR 10 SGV1 *Guideline values for arsenic contamination in soils.*

DEFRA and Environment Agency (2002p) CLR 10 SGV3 *Guideline values for cadmium contamination in soils.*

DEFRA and Environment Agency (2002q) CLR 10 SGV4 *Guideline values for chromium contamination in soils.*

DEFRA and Environment Agency (2002r) CLR 10 SGV5 *Guideline values for inorganic mercury contamination in soils.*

DEFRA and Environment Agency (2002s) CLR 10 SGV7 *Guideline values for nickel contamination in soils.*

DEFRA and Environment Agency (2002t) CLR 10 SGV9 *Guideline values for selenium contamination in soils.*

DEFRA and Environment Agency (2002u) CLR 10 SGV10 *Guideline values for lead contamination in soils.*

DEFRA and Environment Agency (2002v) Contaminated Land Exposure Assessment (CLEA) Model: CLEA 2002 Version 1.3. Developed by the Environment Agency, DEFRA and SEPA. Available as part of CLR 10 from Environment Agency R&D Dissemination Centre, c/o WRc, Frankland Road, Swindon, Wilts SN5 8YF and

DEFRA and Environment Agency (In prep) CLR 10 GV8 *Guideline values for phenol contamination in soils.*

DEFRA and Environment Agency (In prep) CLR 10 GV9 *Guideline values for benzo(a)pyrene contamination in soils, with advice on assessing human health risks from mixtures of polycyclic aromatic hydrocarbons.*

DEFRA and Environment Agency (In prep) CLR 10 GV4 *Guideline values for contamination in soils by inorganic cyanides.*

DEFRA and Environment Agency (In prep) CLR 11 *The Handbook of Model Procedures for the Management of Contaminated Land.*

All DETR and Environment Agency documents available from: Environment Agency Dissemination Centre, c/o WRc, Frankland Road, Swindon, Wilts SN5 8YF and at http://www.defra.gov.uk/environment/landliability/ index.htm

Below is an expected timetable for publication of further TOX and SGV (soil guideline values) papers. These timings may be subject to change.

Contaminant	TOX Papers	SGV Papers
Benzo(a)pyrene	Out now	Summer-03
Inorganic cyanide	Out now	Summer-03
Benzene	Spring-03	Summer-03
Dioxins and dioxin-like PCBs	Spring-03	Summer-03
Phenol	Spring-03	Summer-03
TPH consultation	Spring-03	NA
Toluene	Summer-03	Autumn-03
Ethylbenzene	Summer-03	Autumn-03
Vinyl chloride	Summer-03	Autumn-03
Tetrachloroethane	Summer-03	Autumn-03
PAHs – various	Summer-03	Autumn-03
Atrazine	Summer-03	Mid-04
Azimphos	Summer-03	Mid-04
Dichlorvos	Summer-03	Mid-04
DDT	Summer-03	Mid-04
Carbon tetrachloride	Summer-03	Autumn-03
Xylenes	Summer-03	Autumn-03
Trichloroethene	Autumn-03	Winter-03
Tetrachloroethene	Autumn-03	Winter-03
non-dioxin-like PCBs	Autumn-03	Mid-04
1,2-dichloroethane	Autumn-03	Mid-04
1,1,1-trichloroethane	Autumn-03	Mid-04
Pentachlorophenol	Autumn-03	Mid-04
Chlorotoluenes	Winter-03	Mid-04
TPH consultation response	Winter-03	NA
Endosulfan	Mid-04	Mid-04
Fenitrothion	Mid-04	Mid-04
Malathion	Mid-04	Mid-04

Continued

Contaminant	TOX Papers	SGV Papers
Molybdenum	Mid-04	Mid-04
Thallium	Mid-04	Mid-04
Triflutralin	Mid-04	Late-04
Aldrin-dieldrin	Mid-04	Late-04
Sulfur	Mid-04	Late-04
Vanadium	Mid-04	Late-04
Acetone	Mid-04	Late-04
Beryllium	Late-04	Late-04
Chlorobenzenes	Late-04	Late-04
Trichloromethane	Late-04	Late-04
Explosives – various	Late-04	Late-04
Hexachlorobutadiene	Late-04	Late-04
Organotin compounds	Late-04	Late-04

DETR (1997) CLR 12, '*A Quality Approach for Contaminated Land Consultancy*', DoE (London).

DETR (2000) *Guidelines for Environmental Risk Assessment and Management.* The Stationery Office, P.O. Box 29, Norwich NR3 1GN ISBN 0 11 753551 6.

DoE (1991) Waste Management Paper 27 *The Control of Landfill Gas.* The Stationery Office, London.

DoE (1994a) Contaminated Land Research Report 1, '*A Framework for Assessing the Impact of Contaminated Land on Groundwater and Surface Water*', (Two Volumes), DoE (London).

DoE (1994b) Contaminated Land Research Report 2, '*Guidance on Preliminary Site Inspection of Contaminated Land*', (Two Volumes), DoE (London).

DoE (1994c) Contaminated Land Research Report 3, '*Documentary Research on Industrial Sites*', DoE (London).

DoE (1994d) Contaminated Land Research Report 4, '*Sampling Strategies for Contaminated Land*', DoE (London).

DoE (1994e) Contaminated Land Research Report 5, '*Information Systems for Land Contamination*', DoE (London).

DoE (1995f) Contaminated Land Research Report 6, '*Prioritisation and Categorisation Procedure for Sites which may be Contaminated*', DoE (London).

Environment Agency (1999) *Methodology for the Derivation of Remedial targets for soil and groundwater to protect water resources.* R&D Publication 20, Environment Agency, Bristol.

Environment Agency and NHBC (2000) *Guidance for the Safe Development of Housing on Land Affected by Contamination.* The Stationery Office, London. ISBN 0 11 310177 5.

Environment Agency (2000a) *Assessing Risks to Ecosystems for Land Contamination: Draft Report* Technical Report P338. Available from Environment Agency, Rio House, Waterside Drive, Aztec West, Almondsbury, Bristol BS12 4UD.

Environment Agency (2000) *Risks of Contaminated Land to Buildings, Building Materials and Services. A Literature Review.* R&D Technical Report P331. Available from Environment Agency. R&D Dissemination Centre, c/o WRc, Frankland Road, Swindon, Wilts SN5 8YF. ISBN 1 85705 2471.

Environment Agency (2000) *Secondary Model Procedure for the Development of Appropriate Soil Sampling Strategies for Land Contamination.* R&D Technical Report P5-066/TR. ISBN 1 85705 577 2.

Environment Agency (2000) *Technical Aspects of Site Investigation Volume I (of II) Overview.* R&D Technical Report P5-065/TR. ISBN 1 85705 544 6.

Environment Agency (2000) *Technical Aspects of Site Investigation Volume II (of II) Text Supplements.* R&D Technical Report P5-065/TR. ISBN 1 85705 545 4.

Environment Agency (2001) *Assessment and Management of Risks to Buildings, Building Materials and Services.* R&D Technical Report P5 035/TR/01. Available from Environment Agency. R&D Dissemination Centre, c/o WRc, Frankland Road, Swindon, Wilts SN5 8YF.

Environment Agency (2002c) GasSim Landfill Gas Risk Assessment Tool. http://www.gassim.co.uk.

Environment Agency (2002) *Review of Ecotoxicological and Biological Test Methods for the Assessment of Contaminated Land.* R&D Technical Report P300. Available from Environment Agency Dissemination Centre, c/o WRc, Frankland Road, Swindon SN5 8YF.

Finnamore, J., Barr, D., Weeks, J. & Nathanail, C.P. (2002) *Biological methods of assessment and remediation of contaminated land.* CIRIA Report C575, CIRIA London.

ICRCL (1987) ICRCL 59/83 *Guidance on the Assessment and Redevelopment of Contaminated Land (2nd edition).* Department of the Environment, London.

Nathanail, J., Bardos, P. & Nathanail, C.P. (2002) Contaminated Land Management Ready Reference. EPP Press, London & Land Quality Press, Nottingham.

National Rivers Authority (1994) *Leaching Tests for the Assessment of Contaminated Land: Interim NRA Guidance,* R&D Note 301, NRA (Bristol).

SEPA (1998) *Initial Dilution and Mixing Zones for Discharges from Coastal and Estuarine Outfalls.* Policy No. 28 Version 1 September 1998. Available at http://www.sepa.org.uk.

SEPA (2001) *Water Pollution Arising From Land Containing Chemical Contaminants.* Available free of charge from all SEPA offices. Available at http://www.sepa.org%20.uk/publications/leaflets/cont_land/water_pollution_guide.pdf. Accessed on 5 April 2002.

SNIFFER (1999) *Guidance Manual for the Derivation of Level 1 Numeric Targets for Contaminated Soil to Protect Ecosystems Report II.* Report No SR (99)01 II. Foundation for Water Research, Marlow.

SNIFFER (2000)[1] *Framework for Deriving Numeric Targets to Minimise the Adverse Human Health Effects of Long Term Exposure to Contaminants in Soil.* Report Number SR 99 (02) F. Prepared by Ferguson, C.C., Earl, N.J. & Nathanail, C.P. Foundation for Water Research, Marlow.

SNIFFER (2002) (In prep) Revision to SNIFFER (2000) *Framework for Deriving Numeric Targets to Minimise the Adverse Human Health Effects of Long Term Exposure to Contaminants in Soil.* Ferguson, C.C., Nathanail, C.P., McCaffrey, C., Earl, N.J. & Foster, N.

VROM (1994) *Risk Assessment of Contaminated Soils: Dutch Soil Clean-up Criteria.* Ministry of Public Housing, Physical Planning and Environment, Directorate General of the Environment, Directorate Soil, P.O. Box 30945, 2500 GX The Hague, The Netherlands.

VROM (2000) *Circular on target values and intervention values for soil remediation.* Ministry of Housing, Spatial Planning and Environment, Rijnstraat 8, 2515XP, The Hague, The Netherlands.

Wilson, S.A. & Card, G.B. (1999) Reliability and risk in gas protection design. *Ground Engineering* February 1999, and letters in Ground Engineering (March 1999).

WRc (1996) *Ecological risk assessment for chemicals in the aquatic environment.* SNIFFER report SR 3846/1 prepared by WRc.

Department of the environment industry profiles (47 titles) (various dates)

The DOE Industry Profiles provide developers, local authorities and anyone else interested in contaminated land, with information on the processes, materials and wastes associated with individual industries. They also provide information on the contamination that might be associated with specific industries, factors that affect the likely presence of contamination, the effect of mobility of contaminants and guidance on potential contaminants. They are not definitive studies but they introduce some of the technical considerations that need to be borne in mind at the start of an investigation for possible contamination.

Airports (ISBN 1 85 112289 3).
Animal and animal products processing works (ISBN 1 85112238 9).
Asbestos manufacturing works (ISBN 1 851122311).
Ceramics, cement and asphalt manufacturing works (ISBN 1 85 112290 7).
Chemical works: coatings (paints and printing inks) manufacturing works (ISBN 1 85 112291 5).

Chemical works: cosmetics and toiletries manufacturing works (ISBN 1 85 112292 3).

Chemical works: disinfectants manufacturing works (ISBN 1 85 112293 1).

Chemical works: explosives, propellants and pyrotechnics manufacturing works (ISBN 1 851122370).

Chemical works: fertiliser manufacturing works (ISBN 1 85 112289 3).

Chemical works: fine chemicals manufacturing works (ISBN 1 851122354).

Chemical works: inorganic chemicals manufacturing works (ISBN 1 85 112295 8).

Chemical works: linoleum, vinyl and bitumen-based floor covering manufacturing works (ISBN 1 85 112296 6).

Chemical works: mastics, sealants, adhesives and roofing felt manufacturing works (ISBN 1 85112296 6).

Chemical works: organic chemicals manufacturing works (ISBN 1 85112275 3).

Chemical works: pesticides manufacturing works (ISBN 1 85112274 5).

Chemical works: pharmaceuticals manufacturing works (ISBN 1 85112236 2).

Chemical works: rubber processing works (including works manufacturing tyres or other rubber products) (ISBN 1 851122346).

Chemical works: soap and detergent manufacturing works (ISBN 1 85112276 1).

Dockyards and dockland (ISBN 1 85112298 2).

Engineering works: aircraft manufacturing works (ISBN 1 85112299 0).

Engineering works: electrical and electronic equipment manufacturing works (including works manufacturing equipment containing PCBs) (ISBN 1 85112300 8).

Engineering works: mechanical engineering and ordnance works (ISBN 1 85112233 8).

Engineering works: railway engineering works (ISBN 1 85112254 0).

Engineering works: shipbuilding, repair and shipbreaking (including naval ship-yards) (ISBN 1 85112277 X).

Engineering works: vehicle manufacturing works (ISBN 1 85112301 6).

Gasworks, coke works and other coal carbonisation plants (ISBN 1 85112232 X).

Metal manufacturing, refining and finishing works: electroplating and other metal finishing works (ISBN 1 85112278 8).

Metal manufacturing, refining and finishing works: iron and steelworks (ISBN 1 85112280 X).

Metal manufacturing, refining and finishing works: lead works (ISBN 1 85112230 3).

Metal manufacturing, refining and finishing works: non-ferrous metal works (excluding lead works) (ISBN 1 85112302 4).

Metal manufacturing, refining and finishing works: precious metal recovery works (ISBN 1 85112279 6).

Oil refineries and bulk storage of crude oil and petroleum products (ISBN 1 85112303 2).

Power stations (excluding nuclear power stations) (ISBN 85112281 8).

Pulp and paper manufacturing works (ISBN 1 85112304 0).

Railway land (ISBN 1 85112253 2).

Road vehicle fuelling, service and repair: garages and filling stations (ISBN 1 85112305 9).

Road vehicle fuelling, service and repair: transport and haulage centres (ISBN 1 85112306 7).

Sewage works and sewage farms (ISBN 1 85112282 6).

Textile works and dye works (ISBN 1 85112307 5).

Timber products manufacturing works (ISBN 1 85112308 3).

Timber treatment works (ISBN 1 85112283 4).

Waste recycling, treatment and disposal sites: drum and tank cleaning and recycling plants (ISBN 1 85112309 1).

Waste recycling, treatment and disposal sites: hazardous waste treatment plants (ISBN 1 85112310 5).

Waste recycling, treatment and disposal sites: landfills and other waste treatment or waste disposal sites (ISBN 1 85112311 3).

Waste recycling, treatment and disposal sites: metal recycling sites (ISBN 1 85112229X).

Waste recycling, treatment and disposal sites: solvent recovery works (ISBN 1 85112312 1).

Profile of miscellaneous industries, incorporating: Charcoal works, dry-cleaners, fibreglass and fibreglass resins manufacturing works, glass manufacturing works, photographic processing industry.

Publisher contact details

British Standards Institution, 389 Chiswick High Road, London W4 4AL. http://www.bsi.org.uk.

Building Research Establishment, BRE Bookshop, CRC Limited, 151 Roseberry Avenue, London EC1R 4QX. http://www.bre.co.uk/.

CIRIA, 6 Storey's Gate, Westminster, London SW1P 3AU. www.ciria.org.uk.

DEFRA Publications, c/o IFORCE Ltd, Imber Court Business Park, Orchard Lane, East Molesey, SURREY, KT8 OBZ (Tel 08459-556000; Fax 0208-957-5012; e-mail defra@iforcegroup.com).

DEFRA CLR Reports 7, 8, 9, 10, 9 TOX and 10 SGV reports are available from http://www.defra.gov.uk/environment/landliability/pubs.htm.

Environment Agency R&D Dissemination Centre, c/o WRc, Frankland Road, Blagrove, Swindon, Wilts SN5 8YF. (Tel 01793 865000; Fax 01793 865001); they can also be ordered on line via http://www.webookshop.com/ea/rdreport.nsf.

Land Quality Press, LQM, SCHEME, University of Nottingham, Nottingham
   NG7 2RD. www.lqm.co.uk.
Scottish Executive. Scottish Executive Development Department. Victoria Quay,
   Edinburgh EH6 6QQ. Tel +44 (0)131 556 8400; e-mail (Enquiries)
   ceu@scotland.gov.uk; Fax +44 (0)131 244 8240; http://www.scotland.
   gov.uk/pages/default.aspx.
SEPA. SEPA Corporate Office, Erskine Court, Castle Business Park, STIRLING
   FK9 4TR. Tel: 01786 457700; Fax: 01786 446885. http://www.sepa.org.uk/.

Note

1. The SNIFFER framework is currently being updated. The 2000 edition contains several incompat-
   ibilities with respect to the policy decisions and exposure factors presented in CLR 9 and CLR 10
   (DEFRA and Environment Agency 2002).

Index